# THE ANALYTICAL CHEMIST'S ASSISTANT.

# THE ANALYTICAL CHEMIST'S ASSISTANT:

## A MANUAL OF CHEMICAL ANALYSIS,

BOTH

QUALITATIVE AND QUANTITATIVE

OF

NATURAL AND ARTIFICIAL INORGANIC COMPOUNDS,

TO WHICH ARE APPENDED

THE RULES FOR DETECTING ARSENIC IN A CASE OF POISONING.

BY

FRIEDERICH WOEHLER,

PROFESSOR OF CHEMISTRY IN THE UNIVERSITY OF GOETTINGEN.

Translated from the German,

WITH AN

INTRODUCTION, ILLUSTRATIONS, AND COPIOUS ADDITIONS,

BY

OSCAR M. LIEBER,

AUTHOR OF THE ASSAYER'S GUIDE.

PHILADELPHIA:

HENRY CAREY BAIRD,

1852.

PHILADELPHIA:
T. K. AND P. G. COLLINS, PRINTERS.

TO

FRIEDERICH WOEHLER, DR. PHIL.,

PROFESSOR OF CHEMISTRY IN THE UNIVERSITY OF GOETTINGEN,

AND TO

MICHAEL TUOMEY, ESQ.,

PROFESSOR OF GEOLOGY AND AGRICULTURAL CHEMISTRY IN THE UNIVERSITY OF TUSCALOOSA, ALABAMA,

THIS WORK

IS RESPECTFULLY INSCRIBED,

BY

OSCAR M. LIEBER.

# PREFACE.

Our age is characterized by no feature more distinctly, than by the fact that "knowledge has become dis-aristocratized and labor has become dignified." Science goes hand in hand with the useful arts, and a Liebig does not disdain to teach the ploughman, nor does the farmer reject what science offers. In such an age, it is not only important to teach how each science may be turned to good and great account in the practical pursuits of life; but he too, it is hoped, does some service, who succeeds in making an important branch, or any part of it, more easy of access, or more available to enlarged numbers.

Noticing, in our literature, the want of a treatise which, in a popular style, should give the directions for testing and analyzing the numerous inorganic compounds of nature and of art, I have considered that, by translating professor Woehler's excellent work, I should be placing a serviceable book into the hands of many persons, to whom the original may be unintelligible. The name of the distinguished author will, I trust, be a recommendation to this volume.

The original appeared in Goettingen, in 1849, under the title of *Beispiele zur Uebung in der Analytischen Chemie*, or *Leading Exercises in Analytical Chemistry*, and was intended by Mr. Woehler for the use of the students in his own laboratory. The large amount of practical information however, embraced within so compact a form, should insure it a wider circulation than the author seems to have contemplated, and I now place it before the American public, enlarged by such additions as I have believed it necessary to make, on account of the somewhat altered design and more extensive purpose of the work.

The additions made by me, consist chiefly in an introduction, describing, in terms as brief as it was possible, the more general manipulations necessary in conducting chemical analyses, a description of the modes of analyzing many substances not given in the original, as also the results of the analyses themselves, of nearly all of the compounds spoken of in the work, as ascertained by the best authorities. Besides these, it appeared useful, wherever it was omitted in the original, to add the percentages of the ingredients *sought*, which exist in the compounds *obtained*, in those cases in which it is impossible, or at least productive of loss, to separate the latter into their individual component parts. The invaluable work of Frezenius on quantitative chemical analysis has been the chief book employed in providing them.

For greater convenience, the centigrade degrees were changed into those of Fahrenheit.

In this shape, I trust, the present volume will not only form a useful acquisition to students of chemistry, but also a convenient book of reference for those persons who are not habitually engaged in analyzing chemical compounds, but to whom, nevertheless, it is sometimes an important object to test the quality and value of different substances.

There are minerals and other compounds so similar to those treated of in the work, that no rules for their analyses have been given. A contrary course would have swelled the work to an unnecessary size, caused an equally unnecessary expense, and would have led to tedious repetitions.

In all such cases, it will hardly be necessary to remark that the rules given for similar compounds are also applicable to them.

O. M. L.

COLUMBIA, S. C., *May*, 1852.

# CONTENTS.

PAGE

# INTRODUCTION.

In writing the introduction to this volume, it is not my intention to prepare an elaborate essay on the uses and purposes of Chemistry, or even to give more than a passing explanation of the aim of Analytical Chemistry; as it is to be supposed that those who may find occasion to refer to this manual have already acquired a sufficient insight into the general rules of theoretical chemistry, on which such a vast number of valuable works have been written that it would be superfluous for me to suggest any particular ones. The chief object of this introduction is to enumerate and explain the various general manipulations which are necessary in the course of the analyses. The more especial ones, relating only to individual cases, are explained in their proper places.

Analytical Chemistry is that science, or rather that branch of chemistry, which provides us with the means of ascertaining the different constituents (*elements*, see table of) of a compound, whether natural or artificial. These compound bodies are either organic or inorganic, and this volume only treats of the latter, the most useful, and of which we possess a far greater knowledge than of the former. Chemical analysis is subdivided into Qualitative and Quantitative, the former only developing and separating the different kinds of elementary constituents of a compound, while the latter goes farther, and ascertains the quantity of each. Both require, in their manipulations, the utmost precaution and neatness; and the latter, in particular, needs such extreme care that, in the eyes of the uninitiated, it may seem to approach to pedantry.

The necessary implements in chemical analysis are: A very accurate balance, under a glass case; a Berzelius lamp, a small spirit lamp; various sets of evaporating dishes and of beaker glasses; glass tubes, rods, and funnels, of different lengths and dimensions; some porcelain crucibles, and one of platinum; distilling apparatus, viz., retort and receiver; a chloride of calcium tube, pipettes, a little furnace, and a blowpipe, with the reagents and other articles belonging to it.

The reagents required are of two kinds, liquid and solid. The former are:—

## ACIDS.

| | |
|---|---|
| Hydrochloric, | Hydrocyanic,* |
| Sulphuric, | Acetic, |
| Hydrosulphuric,† | Oxalic, |
| Sulphurous,* | Tartaric, |
| Nitric, | Silico-Fluoric.* |

## BASES.

| | |
|---|---|
| Caustic Potash, | Hydrate of Lime, |
| Caustic Ammonia, | Bromine Water,* |
| Caustic Soda, | Chlorine Water.* |
| Hydrate of Baryta, | |

## SALTS.

| | |
|---|---|
| Carbonate of Soda, | Chloride of Ammonium, |
| Acetate of Soda, | Hydrosulphate of Ammonium, |
| Hypochlorite of Soda, | Tartrate of Ammonia,* |
| Phosphate of Soda, | Chloride of Barium, |
| Pentasulphuret of Potassium,* | Chloride of Platinum, |
| Nitrate of Potash, | Nitrate of Silver, |
| Oxalate of Potash,* | Nitrate of Mercury,* |
| Chromate of Potash, | Acetate of Lead, |
| Chlorate of Potash, | Succinate of Iron. |
| Carbonate of Ammonia, | |

---

* Those marked thus * are either superfluous, or at least necessary only in very few cases; and are, therefore, only required in very small quantities.

† Sulphuretted hydrogen. This is easily decomposed when in solution, and hence the apparatus should be at hand. (See Fig. 1.)

Also,

Alcohol, Distilled Water.

These should be kept in glass vials or bottles, with ground glass stoppers, and carefully labelled.

The solid reagents are:—

Pure Lead, in granules (galvanically precipitated),
Iron Filings,
Saltpetre,
Borax,
Chloride of Calcium,
Cyanide of Potassium,
Sulphite of Soda,
Nitrate of Soda,
Carbonate of Ammonia,
Carbonate of Baryta,
Protosulphuret of Iron,†
Protoxide of Palladium.*

---

† The protosulphuret of iron is employed in making sulphuretted hydrogen, or hydrosulphuric acid gas. It is

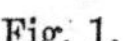

Fig. 1.

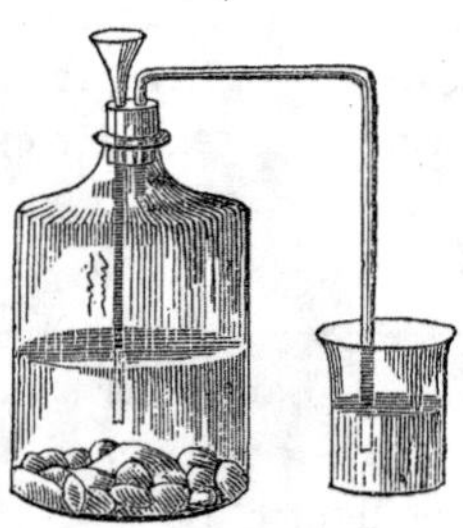

placed in a glass bottle, having two tubes passing through the cork, one to let out the gas, and this should be bent

AND ALSO,

LITMUS PAPER, red and blue, SWEDISH FILTERS.

It will hardly be necessary to remark, that all these should be of the purest, most unadulterated kind. Frezenius, in his famous work on analysis, gives the tests for all, and the modes of purifying them.

The smaller the quantity of the ore, mineral, or artificial compound—which we have under investigation—that may suffice for our purposes, the better in general will the analysis succeed; for the larger the amount, the greater quantity of reagents will be required, and these will be found to be very much in our way. Generally speaking, in the humid or wet process of analysis, one gramme, equal to about seventeen grains (say fifteen or twenty), will be found quite sufficient.

Among the manipulations necessary in the course of the proceedings described in the following pages, the preparation of the various compounds for the action of the different reagents is the first which

---

down, the better to introduce the gas into a beaker glass containing the solution of the substance we wish to precipitate; the other with a funnel top, for the introduction of sulphuric acid and water. To purify the hydrosulphuric acid we pass it through water.

merits our attention. I allude to the reduction of the substances, by *pulverization* or *filing*, to a state in which every portion of the mass is perfectly exposed.

If the mineral, ore, or artificial compound is of such a nature that it is capable of being ground up, an agate mortar is required, or, if it is of too hard a quality, a steel mortar first, to prepare it for the one of agate. The process of pulverization is, perhaps, the most wearisome of all in chemical analyses, particularly when the substance under investigation is one that requires the full extent of this operation to be brought into execution; and, in mere words, it is almost an impossibility to show to what a stretch it is necessary to put the patience and perseverance of the uninitiated, if the object is to attain to the greatest accuracy of which the art is capable. The process must be effected by circular motions, never by perpendicular ones, *i. e.* pounding—or much of the substance will be lost—until the least grittiness is no longer perceptible in the powder. If it be utterly impossible to procure this state in the mortar, as is the case in some few instances, it may become necessary to resort to washing, and thus to remove the grosser particles.

If the substance is malleable, and hence not belonging to the first class, we must have recourse to the file. To this division belong the pure metals or their alloys.

We now come to the process of *digesting*. By this term we designate a treatment of a precipitate or residue, which is necessary in almost every analysis. It is to expose it for some time to a mild heat, with some acid or water, or whatever the peculiar liquid required may be. Generally, a glass vial, to hold the substance, is used for this purpose.

*Filtering* is the next which calls for our attention. This is performed by means of a funnel of glass, the paper being prepared so as to fit into it. For this it is necessary to cut out a circular piece, as shown in Fig. 2, after which the folding will

Fig. 2.

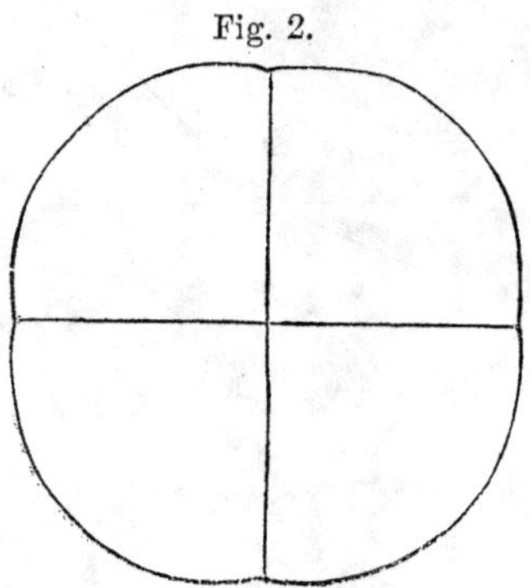

depend on its size, for with the smaller ones we simply fold it according to the breaks in Fig. 2, when it assumes the shape given in Fig. 3. A

Fig. 3.

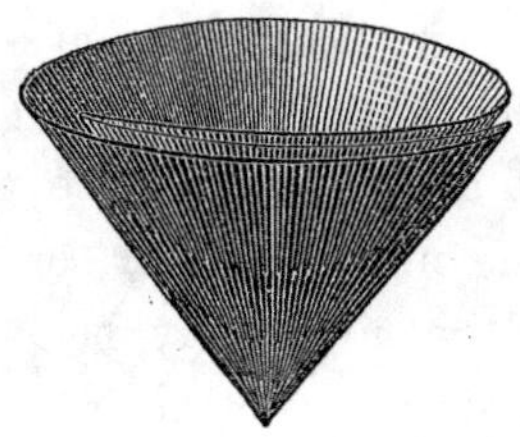

large filter it is necessary to fold in a great many little plaits, Fig. 4, for if it were simply folded as

Fig. 4.

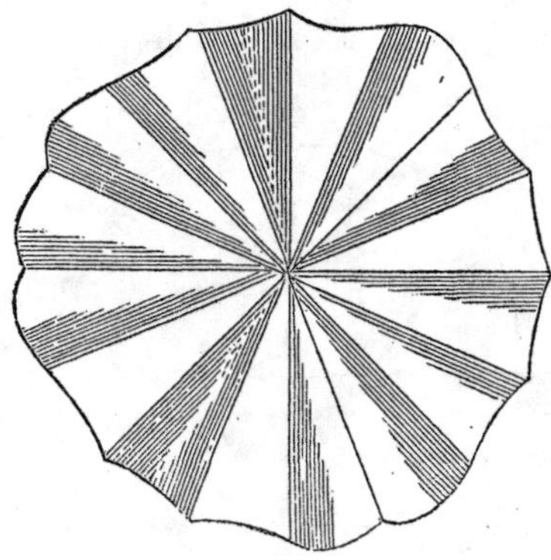

the smaller kind, the weight on the one single unprotected side would be too great, and a break would be the consequence.

We next come to *washing* or *leaching*. In speaking of washing or leaching a precipitate in the filter, it is merely meant that the washing liquid should gradually be poured over it, and allowed slowly to percolate. The washing liquids, as will be seen from the following pages, may be very different. We sometimes use water, sometimes ammonia, sometimes alcohol, and many others. All that is required of a washing liquid is, that it is fully capable of dissolving every particle of the substances contained in the filtrate, while not the slightest portion of the precipitate is soluble in it. We wash either with a pipette or with washing bottles.

The washing bottle, Fig. 5, is a very simple and useful contrivance, consisting, as the figure shows, of a glass bottle or vial with a cork, through which two holes are made for the introduction of two glass tubes. One of these is very short, and does not extend any distance into the bottle. This is for the purpose of blowing in air, and thus forcing out the water or washing liquid through

Fig. 5.

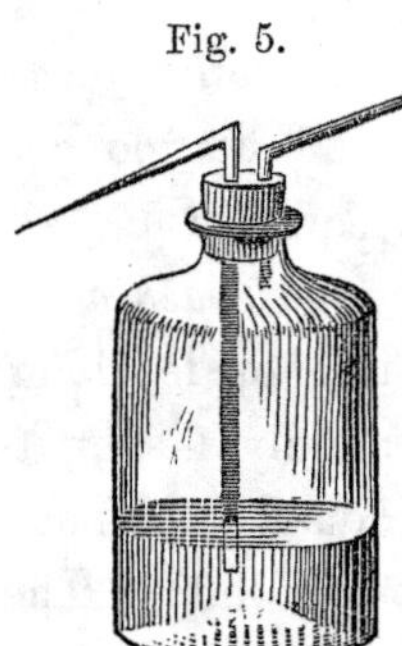

the longer tube, which should be drawn out to a point at the end, outside of the vial.

Fig. 6 represents the self-acting washing bottle,

Fig. 6.

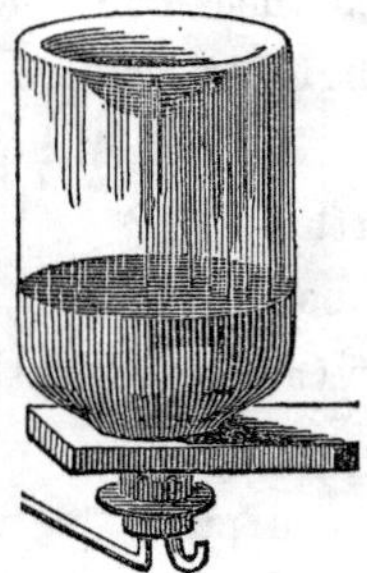

consisting of an inverted vial with two short tubes passing through the cork, and bent as depicted in the diagram. It is used to maintain a perpetual flow or rather dropping of the washing liquid.

*Decanting.*—In the course of an analysis, it is frequently necessary to pour out liquids from one vessel into another. Every one is aware that in a quantitative analysis, where so much may depend upon the most trifling loss, it would be impossible to pour out the contents of a beaker glass or other vessel, without any farther precautions. To avoid the possibility of any of the liquid running down the outside of the glass, some tallow should be rubbed along the outside of that part of it over which we wish to pour the solution. It is neces-

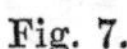

Fig. 7.

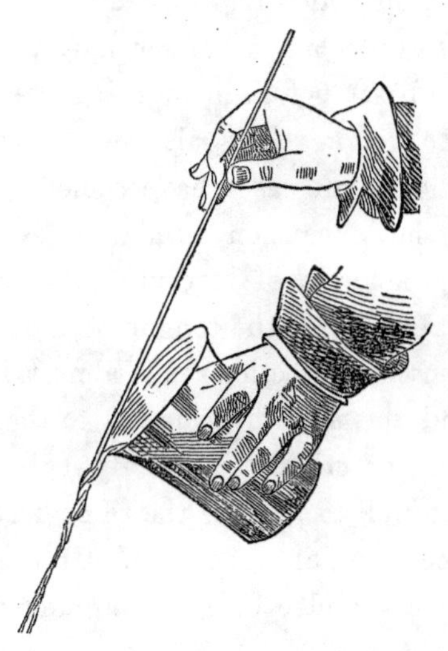

sary to remark here that no large particles of grease should be allowed to remain on the glass, as they might be carried off with the liquid. After making these preparations, a clean glass rod is held immediately over or against the greased portion, so as to shape the course of the solution, while with the other hand we raise the vessel, as illustrated in Fig. 7.

The last mechanical operation in an analysis is the *weighing*. This should be performed with the utmost nicety and precision. To weigh with accuracy it is necessary to dust off every portion of the balance, on which any dust may have settled, with a brush (a common badger hairbrush, cut off even, is the best), before introducing the substance we desire to weigh. It is also necessary that the latter should in all cases have cooled entirely, as otherwise the expansion produced by the heat would be productive of inaccuracy. Ample time should be allowed for the tongue of the balance to settle. Further directions it is almost impossible to give, and the rest must be left to the attention and acquired practice of the analytical chemist.

It will be well to provide the rules here for testing the accuracy of a balance. If the arms of this apparatus, under ordinary circumstances, are perfectly horizontal, and the tongue entirely perpendicular, it does not yet show that the balance is truly correct, but simply that the weight of the arms is well balanced, but it is still possible that one is longer than the other. To ascertain whether or not this is the case, it is necessary to place two equal weights, one on each plate or cup, and then to try again. If, under these circumstances, no alteration takes place, the balance is true in every respect.

With the weighing, all the mechanical operations are completed, and in many cases nothing remains but to make the necessary calculations, as will be shown.

For the convenience of having, as much as possible, all that is necessary in the course of the different analyses embraced within this volume, a table of the combining proportions of the elements is attached at the end of the book. This is taken almost entirely from professor Wöhler's Class-book on Inorganic Chemistry, and has been arranged from the best analyses. The sixty-two elements are placed according to their chemical relationship.

It will be necessary to give some directions for the use of this table. In making the calculation of the percentage amount of the substance *sought*, existing in the compound *found*, we proceed thus: $A$ being the sum of the equivalents of the constituents of the latter, while $X$ is the percentage of the simple or at least less compounded substance which it is our desire to separate by calculation from $A$, and $R$ the equivalent of $X$. We say:—

$$\frac{R \times 100.00}{A} = X.$$

To make it more intelligible, it will in all probability be preferable to give an example.

Pyrophosphate of magnesia is produced by calcining the ammonio-phosphate of magnesia. The formula is $PO_5,2\ (MgO)$ and the equivalent numbers of the elements are:—

| | | |
|---|---|---|
| 2 Mg | = | 316.28 |
| P | = | 392.04 |
| 7 O | = | 700.00 |
| | | 1408.32 |

Now to calculate the magnesium in the compound, we say:—

$$\frac{316.28 \times 100}{1408.32} = 22.44;$$

for the phosphorus:—

$$\frac{392.04 \times 100}{1408.32} = 27.86;$$

and for the oxygen:—

$$\frac{700.00 \times 100}{1408.32} = 49.70;$$

that is to say, the compound in hundred parts consists of:—

| | | |
|---|---|---|
| Magnesium | = | 22.44 |
| Phosphorus | = | 27.86 |
| Oxygen | = | 49.70 |
| | | 100.00 |

The whole of the oxygen is here placed or added together; two atoms were united in the pyrophosphate of magnesia with the magnesium, forming oxide of magnesium or magnesia, while five were combined with the phosphorus in the phosphoric acid.

These calculations are very necessary, as we are not enabled, except in comparatively few cases, to reduce the different substances to their respective, entirely independent forms, by any chemical process, except with great labor and loss. Very frequently, even with these obstacles, it is impossible.

In the above example, the compound has been completely dissected, if I may so express myself, and exposed in its different elementary component parts; but it is necessary to remark here, that, in giving the final results of analysis, it is not common so entirely to separate all the elements, and it was only with a view of making the process or rather calculations perfectly intelligible, that this was done in the above case.

In stating the results respecting the analysis of pyrophosphate of magnesia, we should have proceeded in the following manner: As already mentioned, there were seven atoms of oxygen in the compound, two belonging to the magnesia, and five to the pyrophosphoric acid. Hence we divide the percentage of oxygen in the compound (49.70) by seven, which gives us 7.10 for each atom, and, therefore, we have:—

| | |
|---|---|
| Magnesium | 22.44 |
| Oxygen | 14.20 |
| | 36.64 for magnesia, |

and,

| | |
|---|---|
| Phosphorus | 27.86 |
| Oxygen | 35.50 |
| | 63.36 for pyrophosphoric acid. |

By adopting this plan, the composition of the

substance becomes much more obvious than would be the case if the magnesium, phosphorus, and oxygen were all given in their unconnected state. This fact will probably be yet more easily explained by the following example.

According to Professor Rammelsberg (*Handwörterbuch der Chemischen Mineralogie*,* vol. i. p. 391) the mineral called yenite, or lievrite, consists of:—

| | |
|---|---|
| Silica | = 29.831 |
| Protoxide of iron | = 33.074 |
| Peroxide of iron | = 22.800 |
| Lime | = 12.437 |
| Protoxide of manganese | = 1.505 |
| | 99.647 |

Now it is very evident that had all the elements been separated, in this case, and the oxygen and the iron each been calculated apart, a very imperfect idea of the character of the mineral would have been conveyed, for it is very important that the two different stages of oxidation in the iron should be carefully distinguished.

Before closing these remarks, it will be necessary

* Encyclopedia of Chemical Mineralogy.

to give some explanation of a somewhat different calculation, which it is frequently necessary to make. I allude to those cases where it is required to ascertain the quantity of a certain combination of a substance (in which state the substance was originally contained in the compound under investigation) by calculation, from a certain other combination of the same substance, in which state the latter was produced by the analysis. The following example will serve better to explain this than could be done without any specified case. To avoid repetition, we may employ another compound.

A mineral contains magnesia in the shape of a sulphate, and by the analysis we have procured the pyrophosphate, which, as already mentioned, contains 22.44 per cent. of magnesium, or 36.64 per cent. (more accurately 36.637) of the oxide of magnesium (magnesia), while sulphate of magnesia contains 34.015 per cent. of magnesia.

Let us now suppose that we found the mineral to contain, say ten grains of pyrophosphate of magnesia, or rather that, in the course of the analysis, that amount of that compound was produced; in other words, therefore, that the amount of the mineral employed, contained 3.6637 grains of magnesia, which was originally combined with sulphuric acid, having been in the state of sulphate of magnesia. Therefore, we say, 34.015 parts of magnesia are contained in 100.000 parts of sulphate of magnesia, how much of the sulphate do 3.6637 parts of magnesia make? Or,

$$\frac{100 \times 3.6637}{34.015} = 10.859 \text{ sulphate of magnesia.}$$

In the work of Frezenius, already mentioned in the preface, and in Rammelberg's *Stoechiometrie*, Berlin, 1842, very elaborate treatises on the calculation of analyses have been given, to which I must refer those who may desire further information with regard to it, as it would not come within the compass of the plan of this work to dwell more at large on this subject, though the remarks here made will be sufficient for most practical purposes.

# A MANUAL

OF

# CHEMICAL ANALYSIS.

---

## I.—POTASH AND SODA.

IF it is desirable to ascertain the quantities of each, and not merely to effect an inaccurate separation, it is necessary that both these substances should be present in the state of chlorides. If not originally in that shape, it is only required to neutralize the solution with hydrochloric acid, then to evaporate to dryness, and afterwards heat the residue to redness in a platinum crucible, and weigh, to have the exact weight of the compound. Then slightly moisten it, and add drop by drop an excess of a rather concentrated solution of chloride of platinum. After that the solution, together with the precipitate, should be evaporated to dryness in a water bath, and alcohol poured over the whole, and, after several hours of maceration, the liquid portion, containing the soda, filtered off. The residue, being platino-chloride of potassium, is dried and weighed, the exact weight of the filter at 212° F. being known. This is done after first washing it off with alcohol, and at a temperature of 212° F. 100 parts of this salt are equal to 30.55 parts of chloride of potassium, or 19.33 of potash.

The solution, which contains the platino-chloride of sodium, ought still to be of a strong yellow color. The amount of soda is ascertained by deducting the weight of the potash from the weight of the whole compound.

### II.—AMMONIA.

If a solution contain no other than an ammonia salt, it should be mixed with hydrochloric acid and chloride of platinum, and then slowly evaporated to dryness in a water bath, the residue mixed with alcohol, and the precipitated platino-chloride of ammonia filtered on a weighed filter, washed with alcohol, and dried at 212° F.

If a solution, or a salt, should contain other bases besides the ammonia, it is to be distilled with caustic soda, until all the ammonia is driven off and transferred to the receiver. The ammonia, and the other distilled solution, we condense in diluted muriatic acid.

This solution is then to be evaporated to dryness in a weighed vessel, and the sal-ammoniac weighed as such. Or chloride of platinum is mixed with it, the whole evaporated to dryness, the excess of chloride of platinum washed out with alcohol, and the ammonia-chloride of platinum dried at 212°, and weighed.

### III.—THE ALKALIES AND MAGNESIA.

To separate potash and soda from magnesia in the quantitative analysis all three must be transformed into chlorides. If we have them, as commonly is the case, in a solution, and mixed with salts of ammonia, we evaporate it to dryness, and then heat the residue in a platinum crucible, until the salts of ammonia are evaporated. We may then proceed in three different ways:—

1. Either we place a piece of carbonate of ammonia in the crucible, at the same time heating it to redness and keeping it covered. This we repeat several times, thus changing the chloride of magnesia into magnesia, which then during the solution of the various chlorides of alkalies remains undissolved.

2. Or we add in the crucible, to the mixture of the three chlorides, some pure oxide of mercury, and then heat until all the mercury is evaporated, by which procedure a still more accurate and perfect change of the chlorate of magnesia into magnesia is produced.

3. Or, according to a third method, we digest the three chlorides with carbonate of silver; thus only decomposing the chloride of magnesia; carbonate of magnesia being precipitated. From the filtered mixture of chloride of silver with the excess of carbonate of silver and carbonate of magnesia, we extract the latter with diluted hydrochloric acid.

### IV.—ROCHELLE SALT, OR SALT OF SIEGNETTE.

$$\overline{T},KO,NaO+8aq.$$

By careful heating, the water of crystallization is driven off.

If it is simply desired to establish the fact of potash and soda being present, the concentrated solution should first be mixed with concentrated hydrochloric acid, and then with alcohol. By this process almost all the potash is precipitated as bitartrate of potash, while the soda remains in the solution as chloride of sodium, and may be extracted by evaporation.

To ascertain the quantity of potash and soda present, the salt is first heated to redness, the carbon of the tartaric acid extracted with diluted hydrochloric acid, filtered, and concentrated by evaporation, and then the chlorides of potassium and sodium separated by chloride of platinum, as in Art. I.

### V.—SULPHATE OF MAGNESIA (EPSOM SALTS) AND SULPHATE OF SODA (GLAUBER SALTS), OR SULPHATE OF POTASH.

Several different methods can be resorted to in the analysis of these compounds.

1. According to one, when our intention is to discover an admixture of Glauber salts in the Epsom salts, the salt—we use that which has been exposed to the action of heat, and has *decrepitated*, *i. e.* given off its water of crystallization—is mixed with charcoal powder, and exposed to an intense red heat. If sulphate of soda be present, water will then extract proto-sulphuret of sodium (NaS) which emits sulphuretted hydrogen (hydrosulphuric acid) with hydrochloric acid, common table salt (NaCl) remaining.

2. Magnesia being present, the process should be the following. From the solution of sulphate of magnesia all the sulphuric acid and magnesia is precipitated by hydrate of baryta, $Ba+SO_4$ and MgO (both insoluble substances in water) being formed. From the filtered solution the excess of baryta should be precipitated with carbonate of ammonia. The solution is then to be heated to boiling, filtered and evaporated to dryness, when carbonate of soda remains, the exact amount of which may be ascertained after heating to redness.

From the precipitate of baryta, the magnesia can be extracted with diluted sulphuric acid, and then reprecipitated with ammonia and phosphate of soda, and its weight ascertained. (Compare Arts. VI. and XX.)

3. 100 parts of pure sulphate of magnesia, weighed in a dried state, dissolved in water and mixed with a solution of sal-ammoniac and caustic ammonia, and then precipitated with phosphate of soda, should give 45.5 parts of pyrophosphate of magnesia, weighed after heating, for before that the precipitate is ammonia-phosphate of magnesia. Should it give less, the deficit will be equal to the admixture of sulphate of soda. This way will be found a very practical one, where it is only desira-

ble to ascertain the amount of each of the two *salts* in the compound.

4. If the sulphate of magnesia contain bisulphate of potash, we should produce chloride of potassium, as in the first method of this chapter, with soda, or carbonate of potash, according to the direction given for soda in the second. The first salt is known from its being precipitated by tartaric acid or chloride of platinum, the latter from its absorbing atmospheric moisture, and dissolving, or, as it is termed, being *deliquescent*.

### VI.—CARBONATES OF LIME AND MAGNESIA.

The compound should be dissolved in nitric acid, the solution neutralized with ammonia and the lime precipitated with oxalate of ammonia, or which amounts to the same with oxalic acid and ammonia.

After the solution has perfectly digested in a mild heat, and settled, it is to be filtered, the precipitate washed, dried, heated to redness, moistened with carbonate of ammonia, dried and heated again, and then weighed as carbonate of lime; or else it should be moistened with concentrated sulphuric acid instead of carbonate of ammonia, the excess of acid evaporated, and the lime heated and weighed as sulphate of lime.

The solution which yet contains the magnesia, should receive an excess of caustic ammonia, after which we precipitate with phosphate of soda, wash it with diluted caustic ammonia, dry, heat, and weigh as $(MgO)_2+(PO_5)$. (Compare Art. XX.)

### VII.—CARBONATES OF LIME AND BARYTA.

After dissolving in nitric acid, the solution should be diluted with water, the baryta precipitated with sulphuric acid, and, on allowing time to settle, filtered, washed, dried, and, after burning the filter, heated to redness and weighed. The filtered solution of lime is mixed with an excess of ammonia, and then oxalic acid added as in the last chapter, where this whole process has already been described.

### VIII.—CARBONATES OF BARYTA AND STRONTIA.

The compound is to be dissolved in nitric acid, care being taken not to add an excess of the acid, and the baryta precipitated with silico-fluoric acid. From the filtered solution the strontia is precipitated with an excess of sulphuric acid.

A less accurate method is to dissolve in hydrochloric acid, evaporate to dryness, and to extract the chloride of strontium by alcohol, in which the chloride of barium is hardly soluble.

### IX.—SULPHATES OF BARYTA AND OF STRONTIA.

These two sulphates are among the most insoluble salts, and therefore, before treating them farther, it is necessary to transform them into carbonates, which is effected by boiling the very carefully and finely ground powder of the mineral in a concentrated solution of carbonate of soda. The supernatant liquid should be filtered off from the newly formed carbonates at a boiling heat, after which the latter are to be washed off, still remaining in the filter, with boiling water, when they may be dissolved in nitric acid, and treated as directed in the last chapter.

The reason why the carbonated earths should be separated at a boiling temperature from the solution of sulphate of soda, is, simply, that, if allowed to cool together, the original sulphates will again be formed.

From this it may be easily seen that a more safe method is to heat the original sulphates to redness in a crucible, together with three times their weight of carbonate of soda in its solid form.

### X.—SULPHATE OF BARYTA, OF STRONTIA, AND OF LIME.

To separate these three, it is of course necessary, first to produce the carbonates of baryta and strontia, as given in the last chapter, after which we should proceed according to the rules given in Articles VII. and VIII.

### XI.—CARBONATE OF BARYTA, OF STRONTIA, OF LIME, AND OF MAGNESIA.

The compound should be dissolved in nitric acid and then diluted with water, after which baryta and strontia are precipitated with sulphuric acid, and separated as above. The filtrate contains the lime and magnesia, which are treated as described in Article VI.

### XII.—HYDRATE OF CARBONATE OF LEAD OR PURE WHITE LEAD.

Pure white lead, $PbO+CO_2+aq$ or $3PbO+HO$, is entirely soluble in diluted nitric acid. After drying at 212° it still gives off much water, when heated to redness, and is converted into the yellow oxide. The amount of water may be ascertained by employing a chloride of calcium tube, and the carbonic acid from the loss. Sometimes it also contains basic acetate of lead.

### XIII.—CARBONATE OF LEAD, CARBONATE OF LIME, LIME, AND SULPHATE OF BARYTA (WHITE LEAD OF COMMERCE).

This compound should be treated with nitric acid, by which the sulphate of baryta remains undissolved. From the filtrate, the lead can be precipitated with sulphuric acid, or better with hydrosulphuric acid, and afterwards the lime, as an oxalate. The barytes is a spurious addition, which, on account of its great specific gravity, is not so readily distinguished, as a larger quantity of the lime might be, which is also only added on account of the greater value of the lead.

### XIV.—CARBONATE AND SULPHATE OF LEAD, AND SULPHATE OF BARYTA (WHITE LEAD OF COMMERCE).

The carbonate of lead is dissolved in diluted nitric acid, the two other admixtures remaining. The leached precipitate we digest in a solution of tartaric acid, containing an excess of ammonia, by which process the sulphate of lead is dissolved. From this solution the lead is precipitated with hydro-sulphuric acid or chromate of potash. After this the sulphate of baryta alone remains.

### XV.—ALUM (NATURAL, NATIVE).

$(KO,SO_3+Al_2O_3S_3O_9+24\ aq.)$

From the solution in water the alumina should be precipitated with ammonia, the filtrate evaporated to dryness, and the residue heated to redness in an atmosphere of carbonate of ammonia, to drive off the sulphate of ammonia, when sulphate of potash remains.

The alumina, precipitated in the above manner, yet contains sulphuric acid and potash, from which we can free it, by dissolving in hydrochloric acid, and reprecipitating with ammonia.

For a quantitative analysis, the sulphuric acid contained in the solution of alum should be precipitated with chloride of barium.

The liquid portion, filtered off from the sulphate of baryta, still contains the alumina, together with a probable excess of the baryta, which we previously introduced for the sulphuric acid. Both of these we may precipitate by adding a mixture of caustic and carbonated ammonia.

After applying heat, this precipitate should be filtered off, the solution evaporated to dryness, and the residue heated until all the sal-ammoniac has left. Nothing remains but chloride of potassium.

The precipitate of alumina and carbonate of baryta which we obtained above ought to be dissolved in hydrochloric acid, the solution heated to drive off the carbonic acid gas, and then diluted, when ammonia will precipitate the alumina. This should be filtered off very quickly, allowing as little atmospheric air to act upon it as possible, as otherwise the carbonic acid which might be brought in contact with the solution, would reprecipitate the baryta. After leaching and drying, the precipitate is to be heated to redness and weighed.

The water contained in the alum can be calculated by the loss, or else its amount may be ascertained directly by the application of heat, slowly advanced to redness.

### XVI.—FERRO-AMMONIACAL ALUM.

$$(NH_4O,SO \quad FeO_3,S_3O_3+24 \text{ aq.})$$

By careful heating, the water of crystallization is to be driven off, and its weight ascertained.

Another portion of the mineral should then be dissolved, mixed with hydrochloric acid, and the sulphuric acid precipitated with chloride of barium.

A third quantity should be distilled with an excess of a solution of potash, some hydrochloric acid being placed in the receiver to combine with the ammonia. (See Art. II.)

The precipitated oxide of iron, containing also potash, should be filtered off, washed several times, dissolved in hydrochloric acid, precipitated with ammonia, and, after again filtering and washing, dried, heated to redness, and weighed.

### XVII.—POTASH-CHROME ALUM.

$$KO,SO_2 + \left.\begin{matrix} Cr_2O_3 \\ Al_2O_3 \end{matrix}\right\} S_3,O_3 + 24\ aq.$$

By carefully heating a weighed portion, the water of crystallization is driven off, and the amount ascertained by the loss.

Another weighed portion should be dissolved in water, and from it the sulphuric acid be precipitated with chloride of barium, and this precipitate of sulphate of baryta filtered off, washed, dried, heated, and weighed.

From the solution of a third weighed quantity, alumina and oxide of chrome are precipitated by caustic ammonia, when the solution is nearly at a boiling heat. This double precipitate, after separating it from the liquid portion and washing, should be heated to redness, and their weight ascertained, after which it should be mixed with twice its weight of carbonate of soda, and also a double amount of nitre (saltpetre, nitrate of potash), and in this state exposed to a melting heat in a porcelain crucible. Chromate of potash is thus formed, and may be separated and extracted from the alumina by the application of water. The treatment of the chrome is then the same, as will be explained hereafter in the article on chromate of iron.

The alumina precipitate is yet very alkaline, and, therefore, if it is our intention to weigh it, it ought first to be dissolved in hydrochloric acid, and then reprecipitated with ammonia.

The filtrate procured from the precipitate of alumina and oxide of chromium should be evaporated to dryness, and the saline residue heated with usual care, to drive off the ammoniacal salt, and then the heat slowly increased to redness, while a piece of carbonate of ammonia is held in the crucible. The residue is sulphate of potash.

Another though less precise and accurate method of separating alumina and oxide of chromium, is to dissolve the hydrates in caustic potash, and boil the solution, until the oxide of chrome is precipitated.

XVIII. — SPATHOSE IRON ORE ($FeO+CO_2$), FREQUENTLY ADULTERATED BY THE CARBONATES OF MANGANESE, LIME, AND MAGNESIA.

The powdered mineral should be dissolved in hydrochloric acid, nitric acid being added very gradually during the process. Then the solution being diluted with water, it should be neutralized with carbonate of soda, until it has turned to a reddish-brown color, when a concentrated solution of acetate of soda is added, and the whole heated up to the boiling point. In this manner peroxide of iron, and that alone, is precipitated. The filtrate from this should be neutralized and mixed with hypochlorite of soda, and thus permitted to stand for twenty-four hours, when the manganese will be precipitated in the shape of the hyperoxide or peroxide, having the symbol $MnO_2$. The application of heat transforms it into a combination of the protoxide with the sesquioxide, represented thus: $MnO+M_2O_3$.

The filtrate produced after precipitating the manganese, should be examined for lime and magnesia, as given in Art. V.

Another method of analyzing this mineral is the following:—

The mineral is dissolved as in the first case, and slightly digested with a small excess of carbonate of baryta, by which peroxide of iron is precipitated, and this entirely, but, as a matter of course, together with baryta.

This precipitate is washed, after being filtered, and then dissolved in hydrochloric acid, after which the baryta should be precipitated with sulphuric acid, and then the iron with ammonia.

The filtrate procured above from the precipitate of oxide of iron, and carbonate of baryta, still may contain besides the baryta which was added, protoxide of manganese, lime, and magnesia. The baryta is first to be precipitated with sulphuric acid and the manganese, either as in the first process, or as hydrosulphate of manganese, by adding hydrosulphate of ammonia. Lime and magnesia are treated as in Art. V.

A third plan may be adopted when the percentage of protoxide of manganese and of lime and magnesia is very small. In this case, the acid solution should be diluted with a considerable amount of water, and carbonate of soda gradually added, while constantly stirring, until the reddish-brown color appears, and afterwards more, until all the oxide of iron is precipitated. The other bases remain dissolved in the free carbonic acid, and may be treated as above.

The following are the analyses of spathose iron ores from different localities. According to Berthier a lamellar variety, from Baigony, contains: protoxide of iron, 53.0; carbonic acid, 41.0; oxide of manganese, 0.6; and magnesia, 5.4. Beudant gives the composition of an hexagonal variety, from England, as consisting of protoxide of iron 59.97; carbonic acid, 38.72; oxide of manganese, 0.39; and lime, 0.92. A specimen from the Hartz Mountains, in Germany, according to Klaproth, contained 57.50 per cent. of peroxide of iron, 36.00 of carbonic acid, oxide of manganese 3.30, and lime 1.25. The symbol or formula of a spathose iron ore from Ehrenfriedersdorf in Saxony is $2(MnO+CO_2)+3(FeO+CO_2)$, that of the ore from Autun and Vizille, according to Bertier, is $2(FeO+CO_2)+MgO,CO_2$. It is, therefore, principally $FeO+CO_2=1$ atom of $Fe(=61.37)+1$ atom $CO_2(=38.63)$. Including the admixtures of lime and magnesia, we have the formula.

$$\left.\begin{array}{l}FeO\\MnO\\CaO\\MgO\end{array}\right\}CO_2.$$

### XIX.—ALUMINA AND PEROXIDE OF IRON.

Hydrochloric acid should be used as a solvent. We may then proceed in different ways.

One method is to precipitate both as hydrates by means of ammonia, after which the alumina can be extracted with a solution of caustic potash. This separation is, however, not sure and perfect.

A better plan is to precipitate the iron first, as a sesquisulphuret, by boiling the solution with yellow hydrosulphate of ammonia.

The most accurate way of separating the two ingredients of this compound, is to heat the acid solution to a low boiling, while, to reduce the peroxide of iron to the protoxide, sulphite of soda is added, the solution neutralized with carbonate of soda, and caustic soda in solution mixed with it. The whole is then to be boiled until the precipitate is converted into a black powder.

The supernatant liquid, after having been filtered off, should receive a slight excess of hydrochloric acid, after which we add a small quantity of chlorate of potash, and heat to boiling, thus destroying any organic matter which may have been dissolved from the filter, and which would prevent the perfect precipitation of the alumina, which is afterwards effected by adding ammonia.

### XX.—ALUMINA AND PHOSPHORIC ACID.

Supposing that, as generally is the case, the alumina was precipitated together with the phosphoric acid by ammonia, we would proceed thus:—

First dissolving the precipitate in a solution of caustic soda, we dilute it with water, and precipitate the phosphoric acid with chloride of barium or hydrate of baryta, as phosphate of baryta, after which we add more caustic soda, and then bring the liquid nearly up to the boiling point (say 200° F.), after which we filter it off and examine for alumina, as in the last chapter.

The precipitate of phosphate of baryta should be dissolved in hydrochloric acid, when the baryta is to be precipitated with sulphuric acid, and afterwards the phosphoric acid with sulphate of magnesia and ammonia, as ammoniaco-phosphate of magnesia, characterized by its beautifully crystalline form, when examined under the microscope.* This double phosphate of ammonia and magnesia

* The formula of ammoniaco-phosphate of magnesia is,

$$(NO,H_4,Mg_2O + PO_5) + 12HO.$$

forms slowly, and is remarkable for collecting in particular around the parts of the beaker glass, which may have been touched by a glass rod. The real cause of this phenomenon is not exactly ascertained, but it is probable that the glass rod removes a peculiar oleaginous coating, with which glass is usually covered, and that this induces the salt to settle there easier and sooner than in other places. In a quantitative analysis this should be avoided as much as possible, as it is very difficult afterwards to remove it from the glass. Considerable time should be allowed for the formation of the salt.

Another method for ascertaining the amount of phosphoric acid, is to dissolve the phosphate of baryta in diluted nitric acid, and to reprecipitate it with ammonia. After heating to redness, the precipitate is equal to $Ba_5O,P_2O_5$. (See page 38.)

### XXI.—PHOSPHORIC ACID AND PEROXIDE OF IRON.

If it is merely our desire to ascertain the presence of phosphoric acid in the oxide of iron, and not to acquaint ourselves with the quantity, the oxide should be heated to redness, together with about an equal weight of carbonate of soda, after which the alkali salt ought to be extracted by water, when an excess of hydrochloric acid, and then of carbonate of ammonia is added, after which we are enabled to precipitate the phosphoric acid with sulphate of magnesia.

Another method is to dissolve the oxide of iron in hydrochloric acid, to precipitate with ammonia, and mix and digest the whole mass with hydrosulphate of ammonia, so that the oxide of iron is converted into the sesquisulphuret. The liquid portion containing the phosphoric acid, is then filtered off and treated as previously detailed.

In a very accurate quantitative analysis, when there is but little phosphoric acid and much peroxide of iron present, the following process ought to be adopted:—

Dissolve in hydrochloric acid and keep it at a boiling temperature, at the same time introducing sulphite of soda, until the color of the solution has changed to a light green, in other words, until all the peroxide of iron has been converted into the protoxide. Then boil until the vapors of sulphurous acid are no longer perceptible, after which neutralize with carbonate of soda, and then, for the formation of peroxide of iron, add a very small amount of chlorine water, the requisite quantity of which varies according to the amount of phosphoric acid. After this an excess of acetate of soda is added, which causes a white precipitate of phosphate of iron. More chlorine water is then added, drop by drop, until the solution has changed to a reddish color, after which boil, thus causing the precipitate to collect properly. Then filter and extract the phosphoric acid by treating it with hydrosulphate of ammonia, as above, or else dissolve it in hydrochloric acid, mix with sulphate of soda, add caustic soda in excess, and boil until the precipitate is transformed into black $FeO + Fe_2O_3$, when filter. The phosphoric acid is then precipitated with magnesia, as above.

## XXII.—PHOSPHATES AND CARBONATES OF LIME AND MAGNESIA (BONE-ASH).

One method for the analysis of this compound is to dissolve it in nitric acid and precipitate the phosphates of lime and magnesia with ammonia, the carbonate of lime remaining suspended in the liquid until precipitated with oxalic acid.

The precipitate of the phosphates should be dissolved in the smallest possible quantity of acetic acid, and the solution neutralized with carbonate of soda, when the lime is precipitated with neutral oxalate of potash.

From the filtrate the magnesia is precipitated by adding an excess of ammonia, as ammoniaco-phosphate of magnesia, and from the solution filtered off from this precipitate, the remaining phosphoric acid is precipitated with sulphate of magnesia.

Another method is to dissolve the bone-ash in nitric acid, and neutralize, as far as possible, with carbonate of soda, and add acetate of lead to precipitate the phosphoric acid.

From the filtered and washed phosphate of lead, the phosphoric acid is separated by sulphuric acid or sulphuretted hydrogen.

From the solution of lime and magnesia, the excess of lead which may have been added should be precipitated with sulphuric acid, or sulphuretted hydrogen, after which the lime and magnesia are separated, as in Article VI.

A third method is to digest the bone-ash with an excess of diluted sulphuric acid, then to evaporate the whole almost to dryness, and to extract the phosphoric acid with alcohol.

The residue of sulphates of lime and magnesia

is dissolved in water, and treated as detailed in Article V.

Fluoride of calcium is also, in minute quantity, an ingredient of animal bones, especially of the enamel of the teeth. It is a very remarkable fact that fossil bones contain much more fluoride of calcium than recent ones; in some cases even 10 per cent. Even human bones of an ancient date, as bones from the tombs of Egypt, or from Pompeii, appear to contain more fluoride of calcium than those of the present day. In Article LXXV. directions are given for the treatment of this compound.

### XXIII.—BOG-IRON ORE, BROWN HEMATITE, CONSISTING OF PEROXIDE OF IRON, PROTOXIDE OF MANGANESE, ALUMINA, LIME, MAGNESIA, SILICA, PHOSPHORIC ACID, AND SOMETIMES ARSÉNIC ACID, AND COPPER.

The mineral is dissolved in hydrochloric acid, the mass evaporated to dryness at 212° F., redissolved in warm, dilute, hydrochloric acid, and filtered to remove the sand and liberated silicic acid.

The solution ought then to be boiled with sulphite of soda, to reduce the peroxide of iron to the protoxide, and the arsénic acid to the metal (until no odor is any longer perceptible), and then the arsenic precipitated with hydrosulphuric acid gas, as sesquisulphuret of arsenic, which may sometimes contain copper. (See Article XXIV.)

After removing this precipitate by filtration, the solution should be boiled until all the sulphuretted hydrogen is driven off, when carbonate of soda is added, and the whole boiled with an excess of caustic soda, until the precipitate is converted into a powder.

The solution, after filtering, contains all the alumina and a part of the phosphoric acid, which are separated, as in Article XX.

The precipitate consists of peroxide of iron, protoxide of iron, carbonate of manganese, and the carbonates and phosphates of lime and magnesia. It should be dissolved in hot nitric acid; the solution, so far as possible, neutralized with carbonate of soda, and boiled with acetate of soda. By this process all the oxide of iron and phosphoric acid is precipitated. To separate the two, the precipitate is treated as in Article XXI.

The filtered solution contains the protoxide of manganese and lime, and magnesia; these are separated, as in Article XVIII.

## XXIV.—PHOSPHATE AND ARSENIATE OF LEAD.

This compound should be dissolved in nitric acid, the solution mixed with ammonia, and then with hydrosulphate of ammonia, both in excess, and with them slightly warmed for some time (digested). The sulphuret of lead is then removed by filtration. From the filtrate the sesquisulphuret of arsenic is precipitated by hydrochloric acid, the solution filtered from it concentrated, caustic ammonia, and then sulphate of magnesia added, and thus the phosphoric acid precipitated as ammoniaco-phosphate of magnesia.

It may be well to mention here, that the sesqui-sulphuret of arsenic consists of:—

| | | |
|---|---|---|
| Arsenic | = | 60.95 |
| Sulphur | = | 39.05 |
| | | 100.00 |

## XXV.—SCHWEINFURTH GREEN.

$$CuO,\overline{Ac}^{*} + 3(CuO)AsO_3.$$

Boiling with a solution of caustic potash extracts the acids, red oxide of copper remaining, as one-

* Acetic acid.

third of the arsenious acid is converted into arsénic acid.

By adding an excess of acid, and then carbonate of ammonia and sulphate of magnesia, ammoniaco-phosphate of magnesia is precipitated.

If this paint be macerated with a mixture of concentrated hydrochloric acid and alcohol, a solution of chloride of copper is formed, and the arsenious acid remains as a white powder.

If distilled with diluted sulphuric acid, the acetic acid passes over, and may be brought in contact with a base, and weighed as a salt.

XXVI.—IMPURE BLUE VITRIOL, CONTAINING, BESIDES THE SULPHATE OF COPPER, ALSO THE SULPHATES OF IRON AND ZINC.

After dissolving in water, the solution should be slightly acidulated with a little sulphuric acid, and the copper precipitated with sulphuretted hydrogen, as disulphuret of copper, filtered and washed with a solution of hydrosulphuric acid.

The filtrate ought then to be boiled, hypochlorite of soda being very gradually added to change the protoxide of iron into the peroxide, after which it should be neutralized with carbonate of soda, mixed with a solution of acetate of soda, and boiled, by which the iron alone is precipitated.

From the filtered solution the oxide of zinc is precipitated at a boiling temperature with carbonate of soda.

Another way of separating the oxides of iron and zinc, is to mix their solution with an excess of caustic ammonia, thus precipitating the peroxide of iron alone, after which the zinc is precipitated, either with the sulphuret of an alkali, or by boiling with carbonate of soda.

### XXVII.—THE PEROXIDES OF MANGANESE, IRON, AND ZINC.

The solution, which contains the iron in the shape of peroxide, should be mixed with a sufficient quantity of carbonate of soda to form a stable* precipitate, after which acetate of soda is added, and the whole boiled, all the oxide of iron being thus extracted.

After filtering, the solution is to be acidulated with acetic acid, and the zinc precipitated with hydro-sulphuric acid gas. (See also Art. XXXV.)

---

* If too little is added, the precipitate will redissolve on being shaken.

After neutralizing the filtrate either with hypochlorite of soda, or at a boiling temperature with the carbonate, the manganese may be precipitated as in Art. XVIII.

### XXVIII.—GOLD AND SILVER.

Different processes should be adopted with different percentages of silver.

From an alloy, containing less than 15 per cent. of silver, aqua regia will extract all the gold, leaving chloride of silver. For all the silver to be precipitated, the solution should be diluted. The gold may then be precipitated by oxalic acid, or sulphate of iron. (See Art. XXIX.)

From an alloy, containing more than 80 per cent. of silver, nitric acid will dissolve all the silver, while the gold remains. The silver may then, after filtration, be precipitated with muriatic acid, as chloride of silver, which contains 75.28 per cent. of the pure metal.

From an alloy, containing between 15 and 80 per cent. of silver, it is impossible to extract all the silver with nitric acid, or all the gold with aqua regia, because, in the former case, the silver is not sufficiently exposed, and, in the latter, the gold is very soon covered with a film or coating of chloride of silver. It is therefore necessary to resort to another process. The alloy should be melted down in a porcelain crucible, with its treble weight of pure lead. From this alloy nitric acid will extract all the lead and silver, leaving the pure gold.

After filtering off the solution of the two metals, we may precipitate the silver, either with hydrocyanic acid (prussic acid), or after diluting the filtrate, and boiling slightly, with hydrochloric acid. The cyanide of silver (AgCy) formed by the addition of the first reagent, contains 80.58 per cent. of silver.

### XXIX.—GOLD AND COPPER (COIN, WROUGHT GOLD).

We dissolve in a mixture of nitric and hydrochloric acids; this is aqua regia, and should consist of 1 part of nitric acid to 4 of hydrochloric, until the whole of the alloy is dissolved, when the solution should be warmed with an admixture of oxalic acid, by which all the gold is precipitated in its metallic state.

After washing and drying this precipitate it is to be poured into a porcelain crucible, the filter burnt over it, and the gold together with the ashes heated to redness and weighed.

From the solution filtered off from the gold, the copper may be precipitated either with sulphuretted hydrogen, or at a boiling temperature with caustic potash. For this latter, it should however be diluted. To remove an excess of potash, the precipitate should be repeatedly washed with boiling water, after which it may either be heated to redness and weighed as the protoxide of copper, containing 79.84 per cent. of the metal, or heated with charcoal, and thus reduced to the pure metal.

Another method is to precipitate the gold from the acid solution with a solution of pure sulphate of iron, and afterwards the copper, either with hydro-sulphuric acid gas, or by immersion into the not too acid solution of a piece of polished iron, and application of mild heat. The precipitated copper should be carefully washed and dried, and by heating to redness in an open crucible converted into the protoxide.

### XXX.—SILVER AND COPPER, SOMETIMES WITH GOLD (SILVER COIN AND WROUGHT SILVER).

This alloy should be dissolved in tolerably strong nitric acid, and the silver precipitated from the warm solution, by diluted hydrochloric acid, while stirring the mixture.

The filtered and washed chloride of silver should be perfectly dried, as far as possible removed from the filter, and melted in a porcelain crucible over the spirit-lamp. After this the filter ought to be carefully burnt, the ashes added to the chloride of silver, some drops of nitric acid admixed to oxidize the portion of the silver that may have been reduced by the carbon of the filter, and the whole warmed, when a very few drops of hydrochloric acid are added, to convert the nitrate into chloride. The acid is then to be driven off by evaporation, the chloride of silver melted, and weighed on cooling. The percentage of silver which it contains is given in Art. XXIX.

From the filtered solution, protoxide of copper is precipitated at a boiling temperature by caustic potash, then washed, dried, heated to redness, and weighed after burning the filter separately and with great care. (See also last article.)

If this alloy also contains gold, as is, for instance, the case in the Mexican silver coins, it will remain in the first solution in the shape of a brown powder.

## XXXI.—SILVER AND LEAD.

Four different methods can be adopted in the quantitative analysis of this compound.

The one perhaps most commonly resorted to, is the separation by cupellation.* (See also Art. LXI.)

Another plan is to dissolve the alloy in nitric acid, dilute with much water, heat nearly up to the boiling point, and to precipitate the silver as a chloride. (See Art. XXX.)

The lead is then precipitated from the filtered solution, after cooling, by first neutralizing most of the acid with ammonia, and then introducing hydrosulphuric acid gas. One part of this sulphuret contains 86.61 parts of lead.

A third method is to dissolve both metals in nitric acid, to dilute the solution, and add free hydrocyanic acid, thus precipitating the silver as a cyanide. After the precipitate has settled, and the solution become clear, the former should be separated from the latter by means of

---

* In a little volume of mine, "The Assayer's Guide," a more detailed account of this process is given.—*Translator.*

filtration, the filter having been previously dried and weighed at 248° F. It should then be carefully washed, dried at 248° F., and its weight ascertained, from which that of the filter is deducted. Cyanide of silver contains 80.58 per cent. of silver.

From the filtered solution, the lead can be precipitated either after first neutralizing most of the free acid, by sulphuretted hydrogen, or after concentration, by evaporation with sulphuric acid. Sulphate of lead contains 68.28 parts of lead while the sulphuret has 86.61 per cent.

According to a fourth method, both metals are precipitated from their acid solution by carbonate of soda in a slight excess, after which the precipitate is digested with cyanide of potash, which dissolves the silver as argento-cyanide of potash ($Ag+K,CyO_2$), carbonate of lead remaining undissolved, and this, as it contains soda, should be dissolved in nitric acid, and precipitated with hydrosulphuric acid gas or sulphuric acid.

From the solution of the silver salt, nitric acid will precipitate cyanide of silver.

### XXXII.—QUICKSILVER, OR MERCURY.

From the solution which ought to contain the

mercury as a bichloride, it can be separated, and the quantity ascertained by the following processes:—

The solution is mixed with concentrated chloride of tin, and heated for a few minutes very nearly up to the boiling point, and in this way the quicksilver reduced. After this has collected, the supernatant liquid is poured off, and the metal cleansed by digesting it with concentrated hydrochloric acid, washed, and then dried in a weighed vessel over sulphuric acid.

Another plan is to introduce an excess of hydrosulphuric acid gas, and then to filter off the protosulphuret of mercury thus produced, on a filter, previously dried at 212° F., when the precipitate itself is dried at the same temperature and weighed. Care should be taken not to expose it to a greater heat, as it thoroughly evaporates under those circumstances. It contains 86.21 per cent. of mercury.

We can also procure the same combination of mercury by precipitating with caustic ammonia, and adding colorless sulphuretted hydrate of ammonia, digesting in a close vessel and filtering off very quick.

If the mercury was suspended in the solution in the shape of a protoxide, hydrochloric acid will convert it into the chloride, and as such precipitate it, after which it should be separated by filtering on a weighed filter, and dried at 150° F., for even a red heat will evaporate it completely. This chloride contains 84.95 per cent. of mercury.

If the quicksilver, precipitated by sulphuretted hydrogen, is mixed with sulphurets of other metals whose chlorides are not capable of evaporation, it may be separated from these by heating it slightly in an atmosphere of anhydrous chlorine gas.

From amalgama, the mercury may be separated by heating to redness in a retort and distilling it over, while the other metals remain in their metallic state; or we may oxidize them by heating to redness with exposure to atmospheric action.

If lead is in solution together with the mercury, both metals are precipitated with carbonate of soda, after which they are digested with cyanide of potash, by which the mercury is redissolved. From this solution it should be precipitated with hydrosulphuric acid gas. The carbonate of lead still contains soda, and should therefore be converted into the sulphuret. (See Art. XXXI.)

### XXXIII.—SILVER AND MERCURY.

The most simple, though not the most accurate

method is to heat the alloy to redness, and to calculate the amount of mercury from the loss. Frequently, however, in this process, particles of silver are carried off by the mercurial vapors.

A better plan is to dissolve in nitric acid, and then precipitate the silver with hydrochloric acid, or a solution of table salt, as chloride of silver. If the latter is employed, it is necessary first to admix acetate of soda or ammonia, as, otherwise, some mercury would be precipitated together with the chloride of silver.

The solution, after being filtered off from the precipitate of silver, should be concentrated, and while so doing, strong hydrochloric acid should be added, to destroy the nitric. The mercury is then treated as in the last article.

We may also proceed by precipitating the silver from the nitric solution with prussic acid, and afterwards the mercury with sulphuretted hydrogen, or also in the following manner, by neutralizing the solution with potash, adding an excess of cyanide of potassium, and precipitating cyanide of silver from the solution of the double cyanide by nitric acid, and the mercury afterwards as the protosulphuret.

### XXXIV.—IMPURE CINNABAR, CONSISTING OF PEROXIDE OF MERCURY, BISULPHURET OF MERCURY OR CINNABAR AND RED LEAD.

This is a compound of very frequent occurrence as a paint, as red lead is a common adulteration of cinnabar.

The peroxide of mercury is extracted, together with a portion of the lead, by digesting the paint with diluted nitric acid. The remainder of the lead is left in the shape of the peroxide, a puce-colored insoluble powder, together with the bisulphuret of mercury.

From the solution, the lead is precipitated by sulphuric acid, and afterwards the mercury with sulphuretted hydrogen or an excess of chloride of tin.

From the residue, the lead can be extracted, either by heating with diluted muriatic acid as chloride of lead, or by nitric acid, a few drops of oxalic acid being added.

The cinnabar which, as a matter of course, remains, can only be dissolved in aqua regia, and should then be treated as already described.

## XXXV.—GERMAN SILVER (COPPER, NICKEL, AND ZINC).

Dissolve in nitric acid, evaporate most of the excess of acid, dilute with water, and precipitate the copper with hydrosulphuric acid gas.

The filtrate procured from this, should be concentrated by evaporation, and precipitated with a solution of caustic potash, and with it boiled, which causes the protoxide of nickel to settle, while protoxide of zinc is dissolved. From this solution, the latter is precipitated either by sulphuret of potash, or, after saturating with hydrochloric acid, by carbonate of soda.

By adopting this method, the separation of the nickel and zinc is not very precise, as some protoxide of the latter metal is retained by the protoxide of the former, nor is the following mode more accurate.

It is to melt the mixture of the two oxides with common potash (hydrate of), or to precipitate both metals at a boiling temperature with carbonate of soda, and to digest the protoxide of nickel with hypochlorite of soda, thus converting it into the black sesquioxide, from which the protoxide of zinc is extracted by a solution of caustic potash.

A much more accurate method is to heat the dried mixture of the two oxides to redness, and then to expose them to a high temperature in an atmosphere of hydrogen, as long as water is still formed. The protoxide of zinc is then separated from the nickel, by digesting with a concentrated solution of carbonate of ammonia, contact with the atmospheric air being avoided during the process.

Another plan is to precipitate both oxides at a seething temperature with carbonate of soda, then to wash the two on the filter, and, by adding oxalic acid, to convert them into oxalates, which mass, after this, should be dried, and being placed in a glass tube, running out narrow at one end, heated to redness, by which the nickel alone is reduced. The protoxide of zinc is then extracted as directed in the last method for analysis.

A third way, but which can only be employed when no acid is present after the precipitation, is to dissolve the two oxides, previously precipitated by carbonate of soda at a high temperature, in acetic acid, to add a large surplus of this acid, and introduce sulphuretted hydrogen, by which the zinc alone is thrown down, or we may also add an excess of acetate of soda to the muriatic or nitric solution, and then precipitate as above with hydrosulphuric acid gas. The filtrate should be tested for zinc with hydrosulphate of ammonia.

Zinc should only be weighed as the protoxide, and, to convert it into this, the sulphuret should be heated to redness, first slightly, then more so, until the residue no longer diminishes in weight. Oxide of zinc contains 80.26 per cent. of the metal.

### XXXVI.—CARBONATE OF BISMUTH.

This mineral, almost a pure carbonate, has been analyzed by professor Rammelsberg, of Berlin; it was from a gold mine in Chesterfield County, South Carolina, the only known locality of this occurrence of the metal.* According to his analysis it contained:—

| | |
|---|---|
| Peroxide of Bismuth | 82.63 |
| " Iron | 2.55 |
| Alumina | 1.69 |
| Lime | 0.28 |
| Magnesia | 0.52 |
| Silica | 2.97 |
| Carbonic Acid | 6.02 |
| Water | 3.16 |
| | 99.82 |

* This specimen was presented to professor R. by my-

From this it is evident that the mineral is almost a pure carbonate, consisting of peroxide of bismuth, 4 atoms = 11843.00 or 90.28 per cent.; carbonic acid, 3 atoms = 825.00 or 6.29 per cent.; and water, 4 atoms = 449.92 or 3.43 per cent.

For a qualitative analysis, chromate of potash will precipitate the bismuth alone as a chromate from the muriatic solution. It is then soluble in nitric acid, and may be treated as in Art. XXXVII.

All the other ingredients, except carbonic acid, are precipitated from the solution, as previously described in Art. XXIII.

The whole of the carbonic acid, as a matter of course, was evolved when the mineral was first dissolved, and should have been collected in a chloride of calcium tube. Its behavior with acids soon betrays its presence.

### XXXVII.—BISMUTH AND LEAD.

The compound should be dissolved in nitric acid, an excess of sulphuric acid added, after which the solution should be evaporated until the vapors of the latter acid are perceptible, when some water is

---

self. See also the Analysis in his *Encyc. of Chem. Min. Suppl. IV.*, where he terms it Bismuth Spar.

O. M. L.

added to prevent the concentrated acids from destroying the filtering paper, and the sulphate of lead filtered off immediately, which has then to be treated as already described.

From the filtrate peroxide of bismuth is precipitated by carbonate of ammonia, while the liquid is being gradually heated up to a seething temperature. The filtered, washed, and dried precipitate should, together with the carefully burnt filter, be placed in a porcelain crucible, heated to redness and weighed as pure oxide, which contains 89.87 per cent. of the metal.

### XXXVIII.—BRASS, A MIXTURE OF COPPER AND ZINC, SOMETIMES CONTAINING TIN AND LEAD.

This alloy should be dissolved in nitric acid, then most of the surplus acid driven off by evaporation, water added to dilute the solution, and sulphuretted hydrogen introduced to precipitate the copper. After filtering, and then removing the excess of the gas, carbonate of soda is added to the solution, to precipitate the protoxide of zinc.

If the brass also contained tin, it will remain as peroxide, if lead, sulphuric acid should be added, while the evaporation is going on. On diluting the solution the lead is precipitated as a sulphate.

The best brass contains four parts of copper to one of zinc. Tombac and pinchbeck contain more zinc.

## XXXIX.—BRONZE (COPPER AND TIN), SOMETIMES WITH ZINC, LEAD, AND IRON.

One method is to oxidize with nitric acid, evaporate most of the surplus acid, and dilute with water, after which the undissolved oxide of tin is filtered off. From the solution the copper is precipitated by caustic potash at a boiling temperature.

If the bronze contains zinc, lead, and iron, the lead is precipitated, as in the next article, by sulphuric acid, and the copper with sulphuretted hydrogen. The solution filtered from this sulphuret of copper is mixed with chlorate of potash, and heated, to place the iron in a higher state of oxidation, after which it is precipitated with an excess of caustic ammonia. The oxide of zinc remains in solution, and should be precipitated with hydrosulphate of ammonia.

A more accurate plan is to heat the alloy in an atmosphere of anhydrous chlorine gas, by which process the tin, zinc, and iron are converted into volatile chlorides, and chloride of copper and chloride of lead remain, and are separated as above.

### XL.—SOLDER (TIN AND LEAD).

In analyzing this compound we should oxidize with nitric acid of medium strength, by which the tin remains as an insoluble peroxide, which, after diluting the solution should be filtered off, washed, dried, heated to redness, and weighed. It contains 78.62 per cent. of the metal.

From the filtered solution, the lead is then precipitated with diluted sulphuric acid, and, together with the supernatant liquid, exposed to the action of heat, until the nitric acid has evaporated, and the vapors of the sulphuric acid have become perceptible. The remainder is then diluted with a little water, and the sulphate of lead filtered off on a filter, previously dried at 250° F. and weighed, after which it is washed with alcohol. We may dispense with a weighed filter, if careful to remove the dried precipitate as much as possible from the filter. We must then burn this separately, so that no reduction of lead can take place by means of the carbon of the filter.

XLI.—BISMUTH, LEAD, AND TIN.*

The alloy should be oxidized with nitric acid, when first an excess of ammonia and then a sufficient quantity of yellow hydrosulphate of ammonia ought to be added, and the whole permitted to digest for some time, without the admittance of atmospheric air, until all the tin has been converted into the protosulphuret of that metal, and, as such, dissolved, while the sulphuret of lead and protosulphuret of bismuth remain undissolved.

From the solution, diluted sulphuric acid will pre-

---

* *Newton's fusible metal* is a compound of this kind. It consists, in a hundred parts, of

| | | |
|---|---|---|
| Bismuth | - | 50.00 |
| Lead - | - | 31.25 |
| Tin - | - | 18.75 |
| | | 100.00 |

A very low temperature only is required to fuse it, for it melts below 212° F.

A similar alloy has been employed in some countries, in the place of a safety valve on boilers, in which case the excess of heat, when the steam is condensed to a dangerous extent, causes the alloy to melt, though of course it should not fuse as readily as Newton's alloy, since the minimum temperature of steam is 212° F.

cipitate the tin as protosulphuret. After the evaporation of the hydrosulphuric acid gas, it is to be filtered, washed, dried, and by a carefully and gradually increased heat, first roasted, and then heated to redness, in a porcelain crucible—atmospheric air having access to it—and thus converted into peroxide of tin. During this process, a piece of carbonate of ammonia should be held in the crucible to remove the sulphuric acid. Peroxide of tin contains 80.26 per cent. of tin.

The mixture of the two other sulphurets should be oxidized by nitric acid, and treated as in Art. XXXVII., only that in this case it is not necessary to add sulphuric acid.

### XLII.—ANTIMONY AND ARSENIC.

Different processes are employed in the separation of these two substances. They are generally united in the shape of sulphurets, and, if so, the best plan is intimately to mix one part of the compound with four parts of nitrate of soda and two of anhydrous (decrepitated) carbonate of soda, and thus expose it to heat in a porcelain crucible, until the mass has become oxidized, and has assumed a white color.

After cooling, it should be digested in water, in

which case, this will dissolve the arsenic as arseniate of soda, together with the sulphate of soda, while the antimoniate of soda remains undissolved.

The solution of the former, after being filtered off, is to be acidulated with hydrochloric acid, heated to about 160° F., and, by the introduction of hydrosulphuric acid, the arsenic precipitated as sesquisulphuret of arsenic, which contains 60.95 per cent. of arsenic.

Another plan is to mix the acidulated solution with an excess of ammonia, and to precipitate the arsenic as ammoniaco-arseniate of magnesia, by adding sulphate of magnesia. The precipitate should be leached with ammonia. This double salt possesses, on the whole, the same characteristics as the ammoniaco-phosphate of magnesia, the crystalline deposit is, however, yet more distinct in its individual crystals, when examined under the microscope with transmitted light. (See Art. LV.)

The residue of antimoniate of soda has the formula $NaO,SbO_5$, and from this the antimony should be calculated, antimonious acid containing 80.13 per cent. of the metal. To control this better, the salt should be mixed with more than its weight of sal-ammoniac, and thus exposed to a red heat in a covered platinum crucible, and this process repeated until we find that no change takes place in the weight. In this manner, all the antimony is discharged, and the soda remains as a chloride.

According to a different process, the compound of the two substances, or the mixture of their sulphurets, should be dissolved in aqua regia, when tartaric acid is added—as water might precipitate—and then first an excess of ammonia, and afterwards sulphate of magnesia, thus precipitating ammoniaco-arseniate of magnesia. After carefully heating, this double salt is changed into $Mg_5O+As_2O_5$.

From the filtered solution, after again acidulating, the antimony is precipitated with hydrosulphuric acid.

The sesquisulphuret of antimony contains 72.89 per cent. of antimony.

### XLIII.—ANTIMONY AND LEAD (TYPE-METAL).

The alloy should be reduced to the smallest possible particles, and then oxidized with nitric acid, after which some hydrochloric acid is added, so that the greater part of the antimony is dissolved, and also a part of the lead is precipitated as a chloride. An excess of ammonia is then added, and afterwards a sufficient quantity of hydrosulphate of ammonia, when the whole mass should be digested, with the exclusion of atmospheric air, first slightly warming, but gradually increasing the temperature, until the precipitate of sulphuret of lead has attained a perfectly black color, while the liquid portion still retains a yellow tint.

After cooling entirely the sulphuret of lead is removed by means of a filter previously dried at 250° F., and weighed, and while yet on the filter, washed.

From the filtrate, the sulphuret of antimony is precipitated with diluted sulphuric acid. This product is, however, not pure sesquisulphuret of antimony, but is mixed with the persulphuret and sulphur, and hence this cannot be weighed directly as a means of calculating the amount of antimony in it. It is therefore necessary to filter it off on a filter, dried at 212° to weigh it, and then to take a small weighed portion of it, and mix this very perfectly with two parts by weight of dry carbonate of soda and four parts of nitrate of soda, and placing the mixture in a porcelain crucible over a spirit-lamp, to apply heat until it has assumed a perfectly white color. After cooling, it should be macerated in water, and the insoluble antimoniate of soda removed by filtration and washed.

This residue is $NaO,SbO_5$, and from it the amount of antimony is calculated, first, for the part we examined separately, and then for the whole.

To control the analysis, it is well to examine the solution last obtained, first, by adding hydrochloric acid, and then by chloride of barium, thus extracting the sulphuric acid, and next calculating the amount of sulphur from the sulphate of baryta. This precipitate should be previously washed with great care, by pouring boiling water over it, and allowing it to percolate.

### XLIV.—PEWTER (TIN, ANTIMONY, COPPER, BISMUTH, AND SOMETIMES LEAD).

The pulverization, so to speak, of substances of the nature of the one now under investigation, I may here remark, should be performed with the aid of a good file, or if, as in this peculiar case, the metal or alloy is very soft, it can be effected by cutting or scraping with a penknife. (See also INTRODUCTION.)

After thus reducing the pewter to the shape of filings, nitric acid is poured over it to oxidize the metals, and the copper is thus dissolved. (See Art. XXXIX.) The solution is then to be diluted, and filtered, and from it the copper precipitated at a seething temperature by caustic potash.

The residue, after the whole of the adhering portion of the copper solution has been removed by washing, still contains antimony, tin, bismuth, and also lead, if the latter was contained in the pewter, as is the case with the inferior qualities. Some hydrochloric acid is now to be poured over this mixture of the oxides of the different metals, by which the greater portion of the antimony is dissolved, and the lead converted into chloride of lead, as in the last article. The whole should then be digested, with the addition first of an excess of caustic ammonia, and then of hydrosulphate of ammonia, by which means the tin is dissolved as a protosulphuret, and is suspended in the solution together with the antimony.

The sulphuret of lead, and protosulphuret of bismuth are insoluble in the liquid, and are therefore separated from it by filtration. After carefully washing this precipitate, the separation of the two sulphurets is conducted precisely as directed in Article XLI.

The filtrate, containing tin and antimony in solution, is to be evaporated to dryness, and the residue melted together, with hydrate of soda and some nitrate of soda, as shall be more particularly detailed in Art. XLVI. The separation takes place as given there, and the calculation of the antimony afterwards, as in Art. XLII.

From the stannate of soda the tin is extracted by precipitation from its aqueous solution by hydrosulphate of ammonia and sulphuric acid, and afterwards treated as in Art. XLI.

### XLV.—ARSENIC AND LEAD.

There are different methods by which to proceed in the separation of the two ingredients of this compound, but it is always first necessary to oxidize with nitric acid.

One plan is, then, to evaporate most of the acid; to add an excess of ammonia, and then of hydrosulphate of ammonia, and to let the whole digest a little. Sulphuret of lead is thus precipitated, while the solution contains all the arsenic, which, after filtration, can be thrown down as sesquisulphuret of arsenic by adding hydrochloric acid.

Another method is to add sulphuric acid to the oxidized mixture, to digest for some time, thus to convert all the arsenious acid into arsenic acid, and then, by evaporating the chief part of the surplus sulphuric acid, and adding alcohol, to separate the arsenic acid from the sulphate of lead. By deducting the weight of the pure metal in the latter from the weight of the whole, we have as product the weight of the arsenic.

### XLVI.—ARSENIC, TIN, AND ANTIMONY.

In this analysis, we commence again by oxidizing with nitric acid, and then evaporate to dryness, and melt the residue in a silver or platinum crucible, with an excess of hydrate of soda, and a little nitrate of the same. After cooling, the whole is to be macerated in water, with the aid of heat, and when this again has cooled, the liquid portion containing arseniate and stannate of soda should be separated from the antimoniate of soda by filtration. If this latter should pass through the filter, as is sometimes the case in the course of this process, it should be washed with diluted carbonate of soda, and the amount of antimony ascertained as in Art. XLII.

The solution of stannate and arseniate of soda should be mixed with an excess of nitric acid, evaporated to dryness, and the residue submitted to the action of water. The stannic acid remains insoluble and united, according to the relative quantity of the arsenic, either with the whole, or only a portion of the arsenic acid, while, in the latter case, the remainder of this acid passes over into the solution, and can be precipitated from this by sulphuretted hydrogen, or sulphate of magnesia and ammonia.

The residue containing the stannic acid, and at least a portion of the arsenic is then to be heated in an atmosphere of hydrogen, by which a sublimate of metallic arsenic and reduced arsenic-tin is produced. The latter should be dissolved in aqua regia, the solution placed in Marsh's apparatus (see the Medico-Judicial Process for arsenic at end of vol.), and the hydrogen thus produced permitted to pass through a red-hot glass tube filled with small pieces of copper wire.

### XLVII.—TARTAR EMETIC.

$$\overline{T},KO,Sb_2O_3,+2aq.$$

By heating up to 212° F., and then weighing again, the amount of the water of crystallization is ascertained.

The salt is now to be dissolved in a comparatively large quantity of boiling water, and while the solution is hot, a current of hydrosulphuric acid gas introduced into it until all the antimony is precipitated. Before allowing the solution to cool, the sesqui-sulphuret of antimony is to be removed by filtration, on a dried (at 250°) and weighed filter, the residue washed with hot water, and dried at 250°.

The filtered solution is then to be evaporated to dryness, and the residue of bitartrate of potash heated to redness, the carbon thus produced, extracted with dilute hydrochloric acid, and washed. The filtrate should be evaporated to dryness in a weighed vessel, and the chloride of potash heated to redness, and weighed.

## XLVIII.—NATURAL SULPHURET OF LEAD (GALENA), PbS.

After reducing the mineral to a very fine powder, fuming nitric acid should be poured over it, by warming and digesting with which the sulphuret of lead is entirely converted into the white sulphate.

If the mixture is then diluted with water and filtered, mere traces only of lead will be found in the filtrate. If the ore contained copper, iron, and silver, they will all remain in solution. The two former are detected by ammonia, the latter by hydrochloric acid.

Had we treated the mineral with a weaker nitric

acid, a compound precipitate of sulphate of lead and sulphur would have been produced, while at the same time some nitrate of lead would have been formed, and as such been suspended in the solution. From this, sulphuric acid will precipitate it. Heat, applied to the dried residue, will drive off the sulphur and leave sulphate of lead, which must be added to that produced from the filtrate.

By boiling the sulphate of lead with a solution of carbonate of soda, we convert it into the carbonate, which after washing is entirely soluble in nitric acid.

Sulphate of lead is readily dissolved in tartrate of ammonia, caustic ammonia also being added in excess. From this solution it can be precipitated by hydrosulphate of ammonia as black sulphuret, or by chromate of potash as yellow chromate of lead. Both these reagents will precipitate all the lead.

Galena generally contains from 83 to 85 per cent. of lead, and sulphur from 13 to 16.45 (see next article on calculation of sulphur). Silver is almost always an ingredient of this ore of lead, though never in a regular or constant amount, and therefore cannot be considered as anything more than an admixture.* Iron occurs as a more common impurification than copper, but is by no means as frequent of occurrence as silver.

### XLIX.—SULPHURET OF LEAD AND SESQUISULPHURET OF ANTIMONY.†

Different methods may be resorted to in this analysis.

---

* Almost all lead of commerce contains a smaller or greater percentage of silver; as in most countries, and in our own among the rest, the extraction of the silver would be productive of a greater expense than the amount of this metal would warrant.

† The purest variety of this mineral is the *Plagionite* (from the Greek, πλαγιος, oblique, on account of the form of its crystallization), which, according to Heinrich Rose, consists of

| | |
|---|---|
| Lead | 40.52 |
| Antimony | 37.94 |
| Sulphur | 21.53 |
| | 99.99 |

Another variety is the *Zinkenite,* or *Bisulpho-antimonite of Lead* (of Thompson), which, according to the same authority, contains:—

| | |
|---|---|
| Antimony | 44.39 |
| Lead | 41.84 |
| Sulphur | 22.58 |
| Copper | 0.42 |
| | 99.23[1] |

[1] See Alger's Phillips's *Min.* pp. 525–6.

In the first, the carefully pulverized mineral should be completely oxidized with nitric acid, after which it assumes a white color. The whole ought then to be saturated with caustic potash, and an excess of a solution of pentasulphuret of potassium poured over it, and then submitted to a high temperature. The solution of the sulphur-antimonial salt thus produced should be filtered off immediately, and the residue of sulphuret of lead washed, first with a weak solution of sulphuret of potash and then with pure water.

From the filtrate sulphuret of antimony is precipitated with dilute sulphuric acid, and after allowing it to collect by heating, separated by filtration. (Compare Art. XLIII.)

Another plan is to mix three parts of the mine-

---

A third mineral, which must be mentioned under this head, is the *Bournonite*, which, according to prof. Rammelsberg, has the following composition:—

| | | | | |
|---|---|---|---|---|
| Lead | 6 | Atoms | = 7767.00 | = 41.77 |
| Copper | 6 | " | = 2374.20 | = 12.76 |
| Antimony | 6 | " | = 4838.70 | = 26.01 |
| Sulphur | 18 | " | = 3621.00 | = 19.46 |
| | | | 18600.90 | 100.00[1] |

[1] See Rammelsberg's *Encyc. of Chem. Min. German*, vol. i. p. 123.

ral with three of dry carbonate of soda, and two of sulphur, and then to melt the mixture in a covered porcelain crucible over a spirit-lamp, very gradually increasing the temperature. After cooling, boiling water is poured over the mass, and heated with it until the antimonial salt is dissolved, when we proceed as above.

If there was any sulphuret of copper in the mineral (see Zinkenite and Bournonite) a mixture of the sulphurets of copper and lead will be produced, and should be oxidized with nitric acid, when we should evaporate nearly to dryness, and extract the soluble copper salt by macerating in water, and afterwards washing with the same. Sulphate of lead remains.

To settle quantitatively the amount of the sulphur, a different portion of the compound is used, and one part carefully mixed with two and a half of saltpetre and three of carbonate of potash, and then heated in a porcelain crucible until perfect oxidation has taken place. We ought then to digest with water, filter off the solution of the oxides, wash, add an excess of nitric acid to the alkaline solution, and to precipitate the sulphuric acid with chloride of barium, as sulphate of baryta, one part of which contains 0.13747 of sulphur.

### L.—SESQUISULPHURET OF ANTIMONY AND PROTO-SULPHURET OF IRON.

These two sulphurets are the chief component parts of the mineral called Berthierite, though occasionally there are adulterations in the way of zinc and manganese, as is the case with the variety from Braeunsdorf near Freiberg in Saxony, analyzed by professor Rammelsberg. Berthier's analysis of that from Martouret, in Auvergne, is,

| | |
|---|---|
| Antimony . . . | 61.34 |
| Iron . . . . | 9.85 |
| Sulphur . . . . | 28.81 |
| | 100.00 |

The mineral should be carefully pulverized, and then oxidized with aqua regia, until the residue of sulphur has attained a pure yellow color. Some tartaric acid is then added, and the solution diluted with water, the antimony not being precipitated. After removing the sulphur by means of a weighed filter, washing it, and drying at 212°, the sulphuric acid formed should be precipitated from the filtrate with chloride of barium, and washed on the filter with hot water. After calculating the amount of sulphur in the sulphate of baryta, it must be added to the principal in giving the total.

After removing the surplus of the salt of barium by sulphuric acid and filtering, sulphuretted hydrogen should be conducted into the solution, and thus the antimony precipitated. (For the further treatment, compare Art. XLIII.)

From the filtrate last produced, the iron is precipitated with hydrosulphate of ammonia. After the sesquisulphuret of iron has settled, aided by heat, and the liquid portion lost the greenish tinge, produced by mechanically suspended particles of the precipitate, this should be separated by filtration, washed with hydrosulphuric acid, by means of a washing bottle (see INTRODUCTION), and dried in the funnel. Then it is placed in a porcelain crucible, while yet in the filter, and heated until the paper has been carbonized, when aqua regia is added to extract the iron, which may then be reprecipitated with ammonia, washed, dried, heated to redness, and weighed as peroxide, one part of which contains 0.9 of iron.

## LI.—RED OR RUBY SILVER.

According to professor Wöhler, a dark variety of this mineral, from Mexico, contained :—

| | |
|---|---|
| Silver . . . | 60.2 |
| Antimony . . | 21.8 |
| Sulphur . . . | 18.0 |
| | 100.0* |

The thoroughly pulverized mineral should be exposed to the influence of a slow current of anhydrous chlorine gas in a tube blown into a bulb in the

Fig. 8.

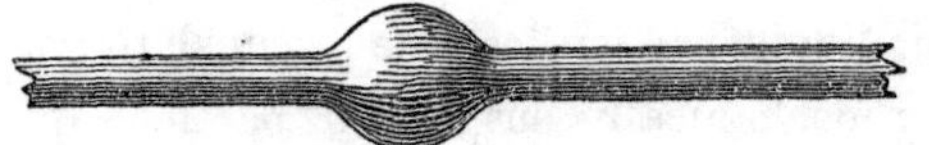

centre, as shown in Fig. 8, the mineral being placed in the latter. By this process, the silver is converted into the not volatile chloride of silver, while the perchlorides of antimony and the dichloride of sulphur, being volatile, are discharged. They may be collected in a suitable apparatus, *e. g.* a receiver, in water acidulated with hydrochloric and some tartaric acid. The action of the chlorine gas takes place even without heat, though it is advisable towards the close of the process, and when the tube is completely filled with the gas, to

* See *Annalen der Pharmacop.* vol. xxvii. p. 157.

apply a spirit-lamp to the bulb, and thus to end the procedure. The bulb will then be found to contain molten chloride of silver, which should be weighed, and, to facilitate this, it would have been well to employ a weighed tube, and then afterwards to deduct its weight, as it will be found very difficult to extract all the chloride of silver from the tube.

The sulphuric acid formed in the liquid contained in the receiver, we precipitate with chloride of barium, and wash the sulphate of baryta with hot water.

The excess of the salt of barium we next throw down with sulphuric acid, filter, and precipitate the antimony with sulphuretted hydrogen from the filtrate. (See Article XLIII.)

### LII.—COPPER PYRITES.

According to Phillips, the calculated composition of this mineral is :—

| | |
|---|---|
| Copper . . . | 34.78 |
| Iron . . . | 30.44 |
| Sulphur . . . | 34.78 |
| | 100.00* |

* See *Ram. Chem. Min.* (German edit.) vol. i. p. 364.

The carefully pulverized mineral should be digested with concentrated nitric acid—concentrated hydrochloric acid being added by degrees—until both copper and iron are dissolved, and a part of the sulphur has been converted into sulphuric acid, and the remainder has collected at the bottom of the vessel, having been melted into a yellow botryoidal mass. This we remove by filtration.

From the diluted solution, the copper is to be precipitated with sulphuretted hydrogen, and the sulphuret of copper washed out with hydrosulphuric acid.

After filtering this off, the solution is heated almost up to a boiling temperature, and, to oxidize the iron still more, some chlorate of potash and hypochlorite of soda added, after which ammonia will throw down all the iron.

A less accurate separation of the ingredients is produced by precipitating the original solution with an excess of caustic ammonia, thus extracting the peroxide of iron, together with a small amount of copper, while the chief part of the latter remains in solution.

### LIII.—SULPHURET OF ZINC, OR ZINC BLENDE.

This mineral almost always possesses adulterations in the shape of iron, and sometimes cadmium and copper.

The analysis of a variety from Ancram, N. Y., by Beck, gave :—

| | |
|---|---|
| Zinc . . . . . | 61.64 |
| Sulphur . . . . | 33.56 |
| Iron . . . . . | 4.30 |
| Gangue . . . . | 0.50 |
| | 100.00 |

while another, by Damour from Nuissière near Beaujeu in the Department du Rhône (France), as given in the *Annales des Mines*, troisième sér. vol. xii. page 245, contained

| | |
|---|---|
| Zinc . . . . . | 62.62 |
| Iron . . . . . | 2.20 |
| Cadmium . . . . | 1.78 |
| Sulphur . . . . | 32.75 |
| | 99.35 |

The well-pulverized mineral should be digested with a mixture of three parts of concentrated muriatic acid, and one of concentrated nitric acid, until all the metals, and the greater part of the sulphur have been oxidized and dissolved, while the rest of the sulphur has collected in pure yellow drops, and all the nitric acid has either been decomposed or discharged.

The solution should then be diluted with water, and an excess of hydrosulphuric acid gas introduced, and thus all the copper and cadmium precipitated.

The precipitate, after being washed with an aqueous solution of the same gas, should be dissolved in concentrated nitric acid, and an excess of carbonate of ammonia added to the solution by which carbonate of cadmium is precipitated, and the oxide of copper dissolved. From this solution a little carbonate of cadmium will yet be thrown down on warming slightly. After filtering, washing, and drying the carbonate of cadmium is converted into the oxide by heating to redness. The latter consists of

| | |
|---|---|
| Cadmium . . . . | 87.45 |
| Oxygen . . . . | 12.55 |
| | 100.00 |

The copper may be treated as directed in the previous analyses.

The filtrate produced after the precipitation with sulphuretted hydrogen, should be heated, and some chlorate of potash and hypochlorite of soda added to convert the prot-oxide into the peroxide of iron.

Caustic ammonia in excess is then employed to precipitate the peroxide of iron, while the zinc remains in solution, and may either be thrown down by hydrosulphate of ammonia, or collected as a residue by adding carbonate of soda, and evaporating to dryness.

The sulphuret of zinc should be washed with hydrosulphuric acid, and, if the analysis is a quantitative one, redissolved in warm concentrated hydrochloric acid, and again precipitated at a seething temperature with carbonate of soda. The precipitate, after washing, drying, and heating to redness, is the pure oxide, and possesses 80.26 per cent. of zinc.

A more precise method for separating the iron and zinc, is to neutralize the solution entirely with ammonia, and to precipitate the peroxide of iron with succinate of ammonia. The precipitate, after heating to redness with access of atmospheric air, leaves peroxide of iron.

For the quantitative calculation of the sulphur, a distinct portion of the mineral should be oxidized as above, the sulphur collected at the bottom, dried at a temperature not exceeding 212°, and weighed. From the solution, the sulphuric acid formed is precipitated with chloride of barium as sulphate of baryta, containing 13.75 per cent. of sulphur, the amount of sulphur contained in it calculated, and added to the rest.

LIV.—PLATINUM, SILVER, COPPER, LEAD, IRON, AND NICKEL MELTED UP TOGETHER AS SULPHURETS.

The mixture, after having been thoroughly pulverized, should be digested with nitric acid, until the process of dissolving has ceased entirely, and all the sulphur has been oxidized and converted into sulphuric acid. The solution is then diluted with water, and carefully poured off from the residue (see Desc. of Manipulations in the INTRODUCTION), which latter should be repeatedly washed.

The residue, consisting of impure sulphate of lead and platinum, is to be digested with basic tartrate of ammonia, and in this manner the sulphate of lead extracted.

The washed residue of platinum is then dissolved in aqua regia, the solution concentrated and mixed with sal-ammoniac, and in this manner ammoniaco-chloride of platinum precipitated. The whole is now evaporated to dryness, and the residue macerated in alcohol, which yet extracts chlorides of copper and of iron.

The ammoniaco-chloride of platinum, on heating to redness, leaves spongy metallic platinum.

Hydrochloric acid is added to the nitric solution, and thus the silver precipitated.

The filtrate procured from this is mixed with the alcoholic solution, and the copper, together with a slight admixture of the lead, extracted with hydrosulphuric acid gas.

The solution filtered off from this is mixed with chlorate of potash, at a boiling temperature, to oxidize the iron yet more, when ammonia in excess will precipitate it, the greater part of the nickel remaining in solution, and capable of being thrown down by hydrosulphate of ammonia.

## LV.—ALLOYS OF COPPER AND NICKEL, ARSENIC AND NICKEL ($Ni_2$ As), AND ORES OF COBALT OR NICKEL, BEING ARSENICAL NICKEL, WITH IMPURITIES OF COBALT, IRON, COPPER, AND BISMUTH.

### 1. *Qualitative Analysis. Preparation of pure Nickel.*

A mixture of one part of arsenical nickel, two parts of nitre, and two of potash, should be heated to redness for some time, and afterwards the arseniate of potash extracted from the mass by means of water, or the arsenical nickel is melted with three times its weight of sulphur and potash, and the potassio-sulphuret of arsenic thus formed extracted by water.

The oxides remaining after the first of these treatments should be dissolved in hydrochloric acid, as well as the sulphurets left after the second process, but with a gradual addition of nitric acid, and with the aid of heat.

The solution of either is to be thoroughly saturated with hydrosulphuric acid gas, and left standing in a closed vessel for twenty-four hours, by which the copper, bismuth, and a small remaining quantity of arsenic are precipitated.

After warming and thus driving off the sulphuretted hydrogen we filter off the liquid portion, heat to a seething temperature, precipitate with carbonate of soda, and carefully wash the precipitate, which contains all the nickel, cobalt, and iron.

While yet in a moist state, a hot, thoroughly saturated solution of oxalic acid is poured over this mixture, which is to be digested with an excess of that reagent, by which all the iron is dissolved, while the cobalt and nickel remain as insoluble oxalates. After filtering and washing, the latter are dissolved by pouring caustic ammonia over them and digesting until nothing of the residue is left.

The blue solution is then set aside in an open vessel until all the free ammonia has evaporated, while at the same time the nickel is thrown down as green ammoniaco-oxalate of nickel (protoxide), the cobalt being retained in solution with a red color.

The filtered and washed nickel salt after heating to redness in a tightly closed (puttied) crucible or glass tube leaves pure metallic nickel.

From the solution of cobalt that metal may be obtained either by evaporating to dryness, and heating the residue to redness, or by boiling with caustic potash until no more ammonia is evolved, or, if the quantity of cobalt is very small, by sulphuret of potash, diluted sulphuric acid being afterwards added so that the sulphuret of cobalt may settle.

From the nickel of commerce the pure metal can be conveniently prepared by the following process, viz.:—

Dissolving in hydrochloric acid with addition of nitric acid, purifying the solution by hydrosulphuric acid gas, and then precipitating cobalt and nickel by adding a boiling, saturated solution of bi-oxalate of potash. The washed precipitate is then dissolved in ammonia, and treated as above.

To extract the iron another plan is to add sal-ammoniac and an excess of ammonia, by which the iron with a small proportion of cobalt and nickel is precipitated, while the greater part of the latter two remains in solution.

2. *Quantitative Analysis.*

We commence by carefully pulverizing the arsenical nickel, after which one part is mixed with two and a half parts of saltpetre, and three of carbonate of soda, and thus exposed to a melting heat, the mass being maintained at a moderate red heat for some time, and after cooling digested with water, and the oxides formed filtered off and washed thoroughly.

The solution which contains all the arsenic as arseniate of alkali, should be saturated with hydrochloric acid, added in excess; evaporated nearly to dryness to drive off the nitric acid, then dissolved in water, heated to about 160° F., and while at that temperature hydrosulphuric acid gas introduced. When, apparently, all the sulphuret of arsenic has been precipitated, we should allow the solution to cool, though the current of sulphuretted hydrogen should not be stopped.

Thus saturated with the gas, the solution is

covered and left for twenty-four hours to settle, after which the precipitate is filtered off on a filter, weighed after drying at 212° ; washed, dried at 212° F. and weighed.

This precipitate is generally a mixture of sesquisulphuret of arsenic and uncombined sulphur. A small portion of it is therefore taken, weighed, and, by warming in aqua regia, completely oxidized. From this solution, diluted with water, the sulphuric acid formed is precipitated with chloride of barium, and from the sulphate of baryta, the sulphur calculated from which the quantity of arsenic contained in the portion used is known, and then, by calculation, its amount in the whole mass ascertained.

Another plan is to precipitate the arsenic acid from the alkaline solution as ammoniaco-arseniate of magnesia. To do this, the solution should be neutralized with hydrochloric acid, mixed first with a solution of sal-ammoniac, then with concentrated caustic ammonia, and afterwards with sulphate of magnesia. After giving the precipitate ample time to settle during the space of twenty-four hours, we should filter, wash with diluted caustic ammonia, and heat the residue, first slightly, but increasing to redness. The product will then possess the formula $Mg_5O+As_2O_5$ (see Art. XLII.), which contains 57.45 per cent. of metallic arsenic.

The amount of arsenic may also be ascertained by the loss.

The oxides should be dissolved in warm concentrated hydrochloric acid; the solution, after evaporating most of the free acid, digested with an excess of carbonate of baryta, by which peroxide of iron alone is thrown down. This precipitate should be removed from the solution by filtration, washed, redissolved in hydrochloric acid, the baryta precipitated with sulphuric acid, and then, after filtering again, the peroxide of iron should be precipitated a second time by ammonia, as an hydrate.

The filtrate from the baryta precipitate is to be saturated with sulphuretted hydrogen, which throws down the copper and bismuth, and these again are separated by dissolving them in nitric acid and adding carbonate of ammonia.

The remaining solution, after extracting the iron and copper, should be heated to a boiling temperature, and, when all the hydrosulphuric acid gas is driven off, the protoxide of cobalt and nickel precipitated with a small excess of caustic soda, a high temperature being kept up. The precipitate, after filtering, should be washed with hot water. The mixture of the two hydrates, while yet in a moist state, should be mixed with diluted prussic acid and solution of potash (*i. e.* cyanide of potassium), and then heated until it is dissolved. The yellowish-red solution should be boiled to drive off the free hydrocyanic acid, and while hot, pure peroxide of mercury, in the finest possible powder produced by washing, added. By this process, all the potassio-cyanide of nickel is decomposed, and all the nickel, partly as oxide, partly as cyanide, precipitated, while the mercury takes its place.

By heating to redness with admission of atmospheric air, the precipitate, after washing, is converted into pure protoxide of nickel, containing 78.68 of the metal.

By drying the mixture of the two oxides, before treating with hydrocyanic acid and potash, and then exposing them to a red heat in a current of hydrogen, and thus reducing them and weighing, we avoid the necessity of ascertaining the amount of cobalt in a direct manner.

If this was omitted, the solution, still containing the cobalt, should be accurately neutralized with nitric acid, and after this a solution of nitrate of mercury, of as neutral a character as possible added, as long as a white precipitate of cyanide of mercury and cobalt is formed. After washing and drying, this should be heated to redness, with access of atmospheric air, by which, while the mercury is discharged, it is converted into the black peroxide of cobalt, which, on account of its varying percentage of oxygen, should, after weighing, be reduced in a current of hydrogen.

If a nickel ore contain lead and sulphur, its analysis is effected according to the process elucidated in Article LI. The compound is decomposed by heating in a current of chlorine gas.

The percentage of sulphur may also be ascertained, by heating to redness a separate portion of the ore with nitre and carbonate of soda, macerating in water, adding an excess of hydrochloric acid, and precipitating the sulphuric acid thus formed with chloride of barium.

LVI.—TIN-WHITE COBALT, $CoAs_2$, or $A_3$, BRIGHT-WHITE COBALT, $CoS_2+CoA_5$, AND COBALTIC FURNACE PRODUCTS, GENERALLY, CONTAINING NICKEL, IRON, COPPER, AND SOMETIMES LEAD AND BISMUTH.

The analyses of these compounds and the preparation from them of pure metallic cobalt, are conducted on precisely the same plan as described in the foregoing article.

Tin-white cobalt occurs in two forms, either one atom of cobalt being combined with two or with three of arsenic. In the following analyses, the first is of an ore containing two atoms of arsenic, from Schneeberg, in Saxony, by Hofmann, the two latter containing three atoms, one a crystalline, the last an amorphous variety.*

| | | | |
|---|---|---|---|
| Arsenic . . | 70.37 | 79.2 | 79.0 |
| Cobalt . . | 13.95 | 18.5 | 19.5 |
| Iron . . . | 11.71 | 1.3 | 1.4 |
| Nickel . . . | 1.79 | —— | —— |
| Bismuth . . | 0.01 | 99.0 | 99.9 |
| Sulphur . . | 0.66 | | |
| Copper . . | 1.39 | | |
| | 99.88 | | |

The bright-white cobalt, according to Rammelsberg,† consists, in its purest shape or by calculation, of

| | | | | |
|---|---|---|---|---|
| Sulphur . | 2 ats. | = 402.33 | = | 19.35 |
| Arsenic . | 2 " | = 940.08 | = | 45.18 |
| Cobalt . | 2 " | = 737.98 | = | 35.47 |
| | | 2080.39 | | 100.00 |

* By Prof. Wöhler (see *Ram. Chem. Min.* vol. ii. pp. 164–5).

† Ibid. vol. i. p. 353.

## LVII.—COBALT AND MANGANESE.

For the purely qualitative separation of the two, we precipitate both the oxides with carbonate of soda, from their solution, redissolve in acetic acid, and introduce hydrosulphuric acid, which will throw down the cobalt, while the manganese is retained in the solution.

For a quantitative analysis, we proceed on the following plan. We first add cyanide of potassium to the acid solution, by which both manganese and cobalt are precipitated, and, by afterwards adding an excess of the same reagent, extract all the cyanide of cobalt and a portion of the cyanide of manganese. The remaining portion of the latter should be removed by filtration, washed and exposed to a red heat, with access of atmospheric air, and when cold, weighed.

The filtrate should be heated up to the boiling, or at least seething temperature, while from time to time a drop of hydrochloric acid is added, which will convert the cyanide of cobalt into the cyanate. More muriatic acid is then added, and the solution boiled again, until at last, with the introduction of still more of the acid, the odor of hydrocyanic acid is no longer perceptible. In this manner, the bicyanide of manganese is converted into the cyanide, and the manganese may be precipitated from the hot solution by carbonate of soda, or caustic soda, while all the cobalt remains suspended in the solution, as potassio-cyanate of cobalt, and may be treated as in Article LV.

The precipitate of manganese, after washing, drying, and heating to redness, should be weighed as above, and the two products added together.

### LVIII.—PLATINUM ORES.

Platinum always occurs native, but as a matter of course with various adulterations, and a great part of these are so intimately allied to that metal, that their perfect separation is almost impossible. The first two of the following analyses are by Berzelius, and the last by Osann, of ore from the Ural Mountains, of the kind worked up in St. Petersburg. The first analysis of Berzelius, is of an ore from Barbacoas, in the province of Antioquia, in Colombia, South America, occurring in coarse grains; the second is of a magnetic ore from Goroblagodat, in the Ural Mountains, and contains no iridium. A portion of the loss in the latter is referable to the osmium.*

| | 1. | 2. | 3. |
|---|---|---|---|
| Platinum . | 84.30 | 86.50 | 80.87 |
| Iron . . . | 5.31 | 8.32 | 10.92 |
| Rhodium . | 3.46 | 1.15 | 4.44 |
| Iridium . . | 1.46 | 1.40 | 0.06 |
| Palladium . | 1.06 | 1.10 | 1.30 |
| Osmium . | 1.03 | — | — |
| Copper . . | 0.74 | 0.45 | 2.30 |
| Quartz . . | 0.60 | — | — |
| Lime . . | 0.12 | — | 0.11 |
| Loss . . | 1.02 | 1.08 | — |
| | 100.00 | 100.00 | 100.00 |

Before proceeding to the analysis, all mechanical admixtures, such as sand and gangue rock, should be separated from the ore. A magnifying glass is very serviceable in this operation. If volatile substances are suspected of being present, it will also be found necessary to heat the mineral to redness and thus to discharge them. A very common admixture is iron in magnetic grains, and

* See Rammelsberg's *Handwörterbuch der Chem. Min.* Supplem. I. for 1843, p. 116.

these are extracted by repeatedly holding a magnet over the grains of platinum ore, and rubbing the adhering particles off, though care should be taken to investigate the quality of the ore beforehand, as frequently magnetic oxide of iron is also contained in the platinum grains, in which case more or less of the platinum would be lost.

The most simple plan is now to dissolve the ore in pure hydrochloric acid mixed with a little nitric acid, in a glass retort or round-bottomed vial. The vessel ought to be placed on a sand-bath, and occasionally, when ebullition takes place, nitric acid added. Should this not entirely decompose the ore, more muriatic and nitric acid should be poured over it, until this is effected. Osmium, iridium, and chromate and titanate of iron (Rutile), &c., remain undissolved, and should be carefully washed. (See next article.)

The acid, distilled over in the dissolving process, and retained in the receiver, should not exhibit any coloring of chloride of platinum, carried over by the force of the vapors. If it does, it must be returned to the retort, and the operation repeated, until the acid retains no coloring matter.

The decomposition of the platinum ores often takes a long time, sometimes even as much as three or four days, and to insure accuracy it is necessary again and again to digest the residue with new acid until the color of the latter is no longer subject to any change.

The different solutions are then to be poured into one vessel and diluted with water. Lime—pulverized and mixed with enough water to give it the consistency of dough—is to be added until the solution has become alkaline, and in this manner iron, rhodium, and the dissolved part of the iridium, &c., are precipitated. The mixture can be treated as in the following article. Platinum alone remains in solution, if the operator is careful not to admit light during the process.

We then filter and leach the precipitate. The filtrate should be acidulated with hydrochloric acid, concentrated, and the platinum precipitated with sal-ammoniac and alcohol as ammoniaco-chloride of platinum, which, after filtering, should be washed with alcohol, placed in a porcelain crucible, and heated, first slightly, but gradually more, until the temperature is increased to a red heat, when a spongy mass of pure metallic platinum is procured.*

* See Bodemann's *Probierkunst*, p. 172, and also technical investigation of the ore furnished in the *Assayer's Guide*, by O. M. L.

## LIX.—PLATINUM RESIDUE, LEFT ON DISSOLVING THE ORE IN AQUA REGIA. ISOLATION AND EXTRACTION OF IRIDIUM, OSMIUM, AND RHUTENIUM.

This residue, which remained undissolved in the process described in the last chapter, consists chiefly of grains and scales of osmium, iridium, iridium in the shape of powder, magnetic iron, titanate of iron, chromate of iron, and generally also of rhutenium. Occasionally, it also contains gold and traces of platinum.

It is first necessary to reduce the coarse grains to as fine a powder as possible, by pounding and grinding with the pestle, the main object being to pulverize the combinations of iron. The whole should then undergo a process of washing (such as is employed in extracting the granules of gold, or gold-dust, from the sand of rivers or their pulverized matrix), by which the chief portion of the osmium-iridium, being present in the shape of granules and minute scales, is mechanically extracted.

The very fine black powder thus also procured should be mixed with about an equal proportion (by size or volume, not weight) of ground, decrepitated table-salt, and the whole then heated to redness in a porcelain or glass tube, in a current of hydrated chlorine gas, until this commences to pass through without being absorbed.

The end of the tube, opposite to the point where the chlorine retort (*A*) is attached, should be fitted into a second retort (*B*), having another opening (*C*), as shown in Fig. 9, from which a tube

Fig. 9.

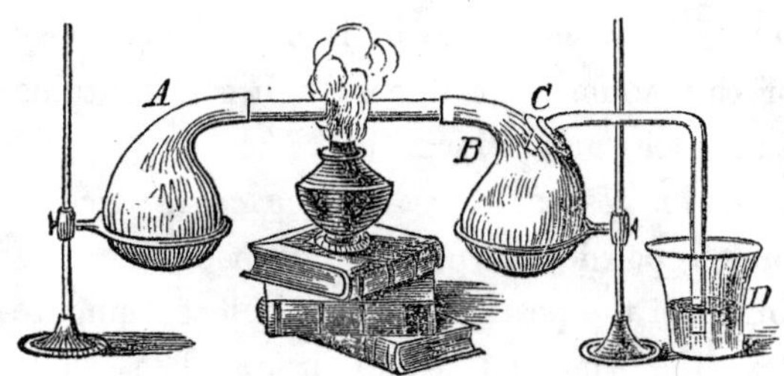


passes out, and conducts the superfluous chlorine gas into alcohol (*D*).

In this manner combinations are formed, consisting of chlorides of iridium and sodium, and chlorides of osmium and sodium. The latter compound is almost entirely decomposed by the moisture of the chlorine gas, and the osmic acid thus produced is sublimated in the retort (*B*), while a part is carried over into the alcohol.

The residue in the tube should, on cooling, be macerated in water, and afterwards washed out on the filter with hot water.

The strongly-colored yellowish-red solution thus filtered from the ferruginous sand, and which contains iridium, should be mixed with concentrated nitric acid, and distilled, by which some of the osmic acid in its aqueous solution is distilled over. The solution of iridium, thoroughly concentrated by this process, ought, before cooling, to be mixed with a saturated solution of sal-ammoniac, by which, as soon as it has become cold, a large portion of the iridium is thrown down as a very dark red, almost black, crystalline mass of ammoniaco-chloride of iridium, which should be filtered and washed several times with a solution of chloride of ammonium. On heating the precipitate to redness, gray, spongy, metallic iridium is left.

The remainder of the solution should be mixed with crystallized carbonate of soda in excess, and this mass then exposed to a moderate red heat in a crucible, and on cooling, leached with hot water, which generally runs off with a yellow color, caused by chromate of soda, the chromic acid originating with the chromate of iron.

The black powder obtained in this manner consists of a compound of sesquioxide of iridium and soda with impurities of peroxide of iron, and sometimes platinum. To reduce it, we heat it to a moderate redness in a current of hydrogen, after which water will extract the caustic soda, while the iron is removed by digesting the residue with concentrated hydrochloric acid. Dilute aqua regia, aided in its action by a mild heat, will generally yet dissolve some platinum (if this was contained in the original residue) which is capable of being precipitated by sal-ammoniac.

The iridium, purified by these several procedures, should be properly washed, and then very strongly pressed and forced to cohere in its different particles. Placing it now in a crucible, it is exposed to a most intense white heat, by which the metal can be forced into a tolerably connected and compact form.

To reduce the osmic acid to metallic osmium, it is necessary to pass it in its vaporous state, together with hydrogen, through a glass tube, heated to redness at one point; or else to mix its solution with formic acid and apply heat; or, according to a third plan, to add ammonia and chloride of ammonium, evaporate to dryness, and heat the whole in a retort until all the sal-ammoniac is discharged, and bluish black osmium is left.

By once only treating the original platinum residue in this manner, it is not yet drained of all the osmium and iridium, more of which is obtained by repeating the process.

If the solution of iridium contain gold, this metal can be precipitated by heating with oxalic acid.

The pure grains and scales of osmium-iridium frequently also containing rhutenium, and procured in the earlier part of this analysis, can be treated like the other portion, after pulverizing in an iron mortar, and then removing the possibly adhering iron by hydrochloric acid. The mass, after heating to redness in chlorine gas, should be dissolved in water, some few drops of ammonia added, and heat applied, by which a mixture of the oxides of rhutenium, osmium, and iridium collects. By distilling this with aqua regia, the osmium passes over as osmic acid. The residue is to be evaporated to dryness, mixed with a compound of potash and nitre (melted up together at a red heat), dissolved in water, when the yellow solution of rhuteniate of potash is to be poured off from the oxide of iridium and carefully neutralized with nitric acid, by which black oxide of rhutenium is thrown down.

A second plan is to melt the entire, though not pulverized, grains with a mixture of equal parts of potash and chloride of potassium, whereupon water will extract osmiate and rhuteniate of potash, while black oxide of iridium, mixed with some unoxidized granules, remains. This solution must not be filtered off.

Rhutenium only occurs at from three to six per cent. in the osmium-iridium, or at from one to one and a half per cent. in the original platinum residue.

### LX.—PLATINUM ORES, CONTAINING PALLADIUM, RHODIUM, AND MERCURY.

The ore is first submitted to the action of nitric acid as in Art. LVIII., and the residue reserved for the processes given in Art. LIX.

The nitric solution should then be galvanically precipitated with a polished bar of zinc, after the ammoniaco-chloride of platinum has been thrown down. A black metallic powder is thus obtained, consisting of platinum, iridium, palladium, rhodium, copper, and frequently mercury.

By applying heat, the mercury is discharged; and its weight ascertained by the loss.

Nitric acid should then be poured over the remainder, and thus palladium and copper obtained in solution. To this we add the impure cyanide of palladium (palladium and copper), procured during the extraction of platinum, and which should have been previously heated to redness; mix with hydrochloric acid and an excess of chloride of potash and evaporate to dryness. Alcohol will then extract the potassio-chloride of copper from the deep red potassio-chloride of palladium, and this latter is then to be reduced to its metallic state after being mixed with sal-ammoniac. Water poured over it will carry off the chloride of potassium.

The metallic residue, from which copper and palladium have been extracted, furnishes the rhodium, and to obtain this separately the powder is melted with bisulphate of potash, as long as this continues to receive a red coloring of potassio-sulphate of rhodium. The salt is then to be dissolved in boiling water, soda in excess added, the solution evaporated to dryness, and then macerated in water, when first hydrochloric acid is added, and decanted, and then the same repeated with water.

Black oxide of rhodium is thus procured as residue, which can readily be reduced in a current of hydrogen.*

The solution may be treated as directed in the last article.

### LXI.—ASSAY OF SILVER ORES BY HEAT.

From argentiferous galena, gray copper ore, copper and iron pyrites and other ores, even when these are mixed with much of the gangue rock, the whole of the silver, concentrated in a little lead, is capable of extraction by the following methods.

A thousand grains of the ore, fully pulverized, are melted together with three hundred grains of saltpetre, and one thousand of litharge. More ore, if used, requires a proportionate increase of the saltpetre and litharge.

Another plan is to melt the ore with from thirty to fifty times its weight of litharge, or one part of ore with three parts of decrepitated (*i. e.* anhydrous) sugar of lead and two parts of potash, under a covering of table-salt.

From the button produced, the silver can be extracted either by driving off the lead in the muffle

* Kœhler's *Chemie in Technischer Beziehung*, pp. 207 and 208.

by cupellation or by the wet process already described.

### LXII.—PEROXIDE OF IRON, AND PROTOXIDE OF IRON, YENITE, MAGNETIC IRON ORE, &C.

For merely technical purposes, it is not always necessary to ascertain the different stages of oxidation in which the iron exists, though even in these it may frequently be an object of consideration to become possessed of the knowledge of this. It were impossible here to enumerate all the minerals which contain these two oxides, and to give their analyses.

Professor Rammelsberg gives the following estimate of the composition of magnetic iron ore, calculated from the formula to which the analyses have been reduced :—

| | | | |
|---|---|---|---|
| Peroxide of iron, | 69.02 | giving iron | 71.78 |
| Protoxide of iron, | 30.98 | and oxygen | 28.22 |
| | 100.00 | | 100.00 |

The mineral termed Yenite, is another which belongs to this class. Its composition is given in the INTRODUCTION.

1. *Modes of determining the whole metallic iron.*

The first plan is to dissolve the substance in hydrochloric acid, and to heat the solution with some chlorate of potash, to convert the chloride into chlorate, and then to precipitate the peroxide of iron with ammonia, wash, heat to redness, and weigh. This process is, however, only applicable, when no alumina, phosphoric, or arsenic acid is present.

Another method, for which we are indebted to Fuchs, is to remove all the free chlorine by boiling from the solution in muriatic acid with chlorate of potash, and then to insert a weighed bright strip of copper, and to boil until the yellow color, produced by the formation of chloride of copper, has changed to the bluish green of the salts of protoxide of iron. The solution is then permitted to cool in a closed vessel, when it should be decanted from the copper, the latter quickly and thoroughly washed off, dried, weighed, and the loss calculated. The equivalent of copper (395.60) is to that of iron (350.53) as the dissolved portion of the copper to the quantity of iron contained in the solution.

2. *Mode of determining the amount of peroxide and protoxide of iron separately.*

The mineral should be dissolved in an excess of

concentrated nitric acid, a closed glass vessel, filled with carbonic acid gas, being employed. Water is then introduced until the whole room, unoccupied by the solution, is filled up, when a weighed strip of copper is inserted, the stopper or cover of the vial or other vessel employed replaced, and this vessel itself set in a basin or evaporating dish containing water, which we gradually heat up to a seething temperature. We then proceed as above, for it is necessary first to know the whole amount of the iron, thus to possess a controlling proof for the accuracy of the remainder of the analysis.

We must next ascertain the amount of peroxide of iron in the compound. To do this, another well-ground portion should be weighed off, and placed in a vial, possessing a cork, through which three tubes are inserted; one having a funnel top, and reaching to the bottom, a second being for the introduction of carbonic acid gas, and the third for the vapors, &c., to pass off (see Fig. 10). Carbonic acid is then forced into the vial, until it is filled, and, after that, hydrochloric acid poured in through the funnel. The dissolving process should be aided by warmth, while the current of gas is kept up throughout. After that, water (previously boiled out) is added, by means of the funnel, to dilute the solution, and then carbonate of baryta, which has been mixed with water until it resembles

Fig. 10.

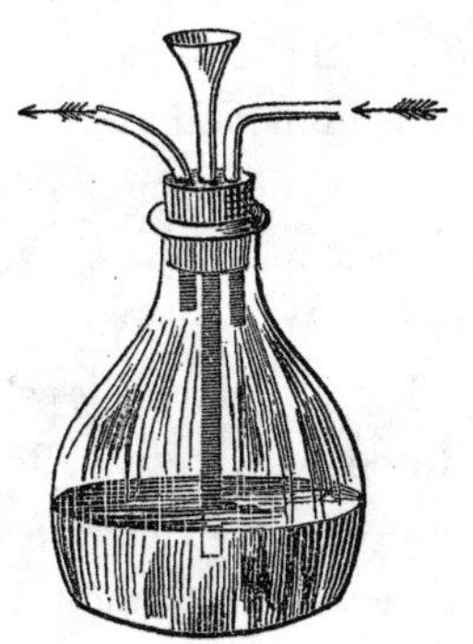

milk. Under these circumstances, by the aid of mild digestion, all the peroxide of iron is thrown down, while the protoxide remains in solution. After the precipitate has settled, and the liquid portion become perfectly clear, the latter should be decanted through the third tube, and this repeated with great care after washing the residue. The latter is then quickly to be placed in a filter, and with equal rapidity leached, by means of a washing bottle, containing previously boiled, but now cold water. The access of atmospheric air must be scrupulously avoided, by placing a plate of glass over the funnel.

The ferruginous residue should then be dissolved in muriatic acid, and first the baryta precipitated with sulphuric acid, and then the peroxide of iron with ammonia.

The liquid portion, which contains the protoxide of iron, should be mixed with a little nitric acid, then concentrated by evaporation, and the baryta and iron precipitated as above. It must be remembered that the iron precipitate is, however, the peroxide, and that it is necessary to calculate the original protoxide, from which it was formed. The peroxide contains only one atom of oxygen to one of iron, while in the protoxide, we have three atoms of the former to two of the latter.

The protoxide of iron contained in the mineral, may also be calculated by deducting the peroxide from the whole amount of iron, ascertained by Fuchs's process. To do this, it is necessary to proceed in the following manner :—

Considering $A$ as the sum total of the metallic iron, $B$ as the peroxide of iron found, $X$ as the protoxide (sought), $b$ as the metallic iron in a hundred parts of the peroxide, and $c$ as the metal in a hundred parts of the protoxide, and $d$ as the metal in the protoxide, ascertained by calculation, and $e$ as the iron in the peroxide; then

$$\frac{B \times 100}{b} = e \text{ (sought.)}$$

From $A$
Deduct $e$

Leaves $d$ (sought.)

$$\frac{100 \times d}{c} = X \text{ (sought.)}$$

### LXIII.—ASSAY OF IRON ORES BY HEAT.

The humid process has already been given in the first part of the last article. The assay by heat is performed in a charcoal crucible, *i. e.* a Hessian crucible lined with charcoal powder.

The carefully pulverized and weighed iron ore, roasted or not, accordingly as it does or does not contain sulphur, should be mixed with calcined borax, and exposed to the most severe heat for an hour, in a tightly closed (with putty) charcoal crucible in a well-ventilated draught, or bellows furnace.

The quantity of borax varies according to the quality of the ore. The more foreign admixtures there exist, the more of the flux is required. Ten parts (about 150 grains are commonly used) of the iron ore require at least three parts of borax, and most generally ten.

The iron thus produced is not pure, but equal in quality to the pig metal (crude iron), produced from the same ore in the high furnace. For its further investigation, see next article.

### LXIV.—CRUDE IRON, DETECTION AND ISOLATION OF ADULTERATIONS.

In detecting and calculating the amount of each of the foreign substances contained in the pig iron, it is best in most cases to employ different portions of the latter.

1. *Carbon.*—The whole quantity of this is discovered and established, by placing the properly filed iron, as in organic analyses, in a small porcelain vessel (shaped like a canoe, and moving with ease in a glass or porcelain tube), and then igniting it in a tube, with a slow current of oxygen, applying heat at the spot where the little vessel is put. The carbonic acid gas, thus evolved, is condensed in a weighed potash apparatus, attached at one end of the tube. This apparatus is then weighed a second time, and the increase in weight will be equal to the amount of carbonic

acid produced from the carbon. One hundred parts of the former contain 27.273 of the latter.

Another plan is the following. A second weighed quantity of iron filings are to be dissolved in diluted sulphuric acid, during which process the chemically combined carbon passes off as carburetted hydrogen, while that existing in the shape of graphite is retained. The gas may be conducted through a solution of acetate of lead, and by the precipitation of sulphuret of lead the presence of sulphur detected.

The insoluble residue should be washed carefully, dried at about 400°, and, as above, ignited in oxygen. From the amount of carbonic acid, that of the graphite is calculated.

2. *Silicon.*—The residue of the first test for carbon, which contains all the silicon as silicic acid, should be dissolved in concentrated hydrochloric acid, the solution evaporated to dryness in a water bath, the remaining mass digested with diluted hydrochloric acid, and the silicic acid removed by filtration. One part of this acid contains 0.48036 of silicon.

3. *Phosphorus.*—The solution filtered off from the silicic acid is treated, for the detection of phosphoric acid, as in the second process in Art. XXI.

If the iron contain arsenic, this will be precipitated with the former in the shape of arsénic acid.

A less accurate plan for the quantitative detection of phosphorus, is to heat to redness one part of the iron filings with one part of carbonate of soda and two of saltpetre; then to macerate the mixture in water, to saturate the solution first with muriatic acid and then with caustic ammonia, and to add sulphate of magnesia.

4. *Arsenic.*—For the mere qualitative detection of this substance, we dissolve the pig iron in diluted sulphuric acid, and digest the filtered black residue with hydrosulphate of ammonium. From this filtered solution diluted sulphuric acid will precipitate sesquisulphuret of arsenic. The precipitate we dissolve in aqua regia, evaporate off the nitric acid, and reduce the arsenic in Marsh's apparatus. (*See* Medico-Judicial Process, at the end of this work.)

The solution of iron, filtered off from the black residue, also contains arsenic, to obtain which we neutralize with carbonate of soda, add a few drops of chloride of iron, and then acetate of soda, by which arseniate of iron is thrown down. This is easily decomposed by hydrosulphate of ammonium.

For the quantitative determination of the arsenic we should dissolve the crude iron in muriatic acid, gradually adding nitric acid; and then, after filtering the solution, and separating it thus from the carbon, heat it with sulphurous acid, until all the protochloride is converted into perchloride, and all the sulphurous acid is discharged. Sulphuretted hydrogen is then introduced in excess. After remaining saturated with this for twenty-four hours in a closed vessel, the gas is permitted to evaporate, and the solution is filtered off from the precipitate. (See below.)

5. *Copper.*—This will be found in the sulphuretted hydrogen precipitate. After drying, the latter should be distilled in a tube, sulphuret of copper remaining. Another plan is to extract the sesquisulphuret of arsenic with caustic potash, though the most accurate way is to do so with sulphuret of potash.

6. *Manganese.*—The solution filtered off from the sulphuretted hydrogen precipitate of arsenic is heated very nearly to boiling, and chlorate of potash or hypochlorite of soda added to convert the protoxide of iron into the peroxide. After this, iron and manganese are separated, as in Art. XVIII., by bicarbonate of soda.

7. *Alumina.*—The alumina is contained in the

iron precipitate just produced, and can be extracted from it as in Art. XIX.

8. *Magnesium and Calcium* remained with the protoxide of manganese in the solution, precipitated by bicarbonate of soda.

9. *Chromium and Vanadium.*—A large quantity of the iron filings are heated to redness with two parts of nitre and one of carbonate of soda; the mass macerated in water and treated as will be shown in another article. Phosphoric and arsénic acids may also be sought for together with these. A more certain and sure method is to employ for this part of the analysis the carbonaceous residue of the solution of a very large quantity of iron in diluted sulphuric acid.

10. *Molybdenum* is sometimes extracted together with the arsenic from the black carbonaceous residue, by hydrosulphate of ammonia, and from the solution (in acid) precipitated with the sesqui sulphuret of arsenic. By distilling this in a tube the arsenic is discharged, while sulphuret of molybdenum remains.

11. *Sulphur.*—The approximate amount of this substance we can ascertain from the hydrosulphuric acid when dissolving the iron in sulphuric acid as in 1. In 3 and 6 we have it in solution, and can

precipitate with chloride of barium. A third method is to dissolve a large quantity of iron in aqua regia, and to precipitate the sulphuric acid formed from the diluted solution with chloride of barium.

12. *Nickel and Cobalt* we may discover by elevating the state of oxidation of the solution, from which the copper was extracted by sulphuretted hydrogen, and precipitating the iron by digestion with carbonate of baryta, after which we are enabled to precipitate nickel and cobalt by hydrosulphate of ammonia.

For the detection of most foreign admixtures it is advisable to use the black residue, produced when dissolving the iron in diluted sulphuric acid. Of this it is easy to procure a large quantity. It contains silica, carbon, carburetted iron, phosphuretted iron, arseniuretted iron, and chromate and vanadiate of iron, molybdenum, &c.

To procure this residue so that it contains the whole amount of carbon (phosphorus, arsenic, chromium, &c.) in the iron, it is necessary to digest the iron filings with a solution of chloride of copper, by which all the free, uncombined iron dissolves, copper taking its place. After decanting the solution, it should be digested with a solution of chloride of iron, access of air being avoided, and in this manner all the copper extracted.

Potash, digested with this residue, will extract a newly-formed humic compound; and besides this, phosphoric, arsénic, and silicic acids. Nearly the whole amount of the latter may thus be obtained.

The residue may also be examined by first heating to redness in chlorine gas.

## LXV.—METEORIC IRON AND METEORIC STONES.

Nickel is an invariable ingredient of meteoric iron, and it is to this that in a great measure its remarkable hardness is referable.

An analysis by Professor B. Silliman, of some of this mineral from Texas, gave the following result. A residue was left on dissolving in hydrochloric acid:—

| | |
|---|---|
| Iron . . . . . . | 90.911 |
| Nickel . . . . . . | 8.462 |
| Residue . . . . . | 0.500 |
| | 99.873 |

The residue was almost entirely soluble in aqua regia, while insoluble graphite remained. The contents were:—

| | |
|---|---|
| Iron . . . . . . . | 31.2 |
| Nickel . . . . . . | 42.8 |
| Phosphorus . . . . . | 4.0 |
| Carbon . . . . . . | 5.0 |
| Antimony? } Copper } . . . . . | 9.3 |

A meteoric stone from Tulbagh in the Cape Colony, which fell on the 13th Oct., 1838, contained, according to Faraday, besides traces of cobalt and soda—

| | |
|---|---|
| Silica . . . . . . | 28.90 |
| Protoxide of iron . . . . | 33.22 |
| Magnesia . . . . . . | 19.20 |
| Alumina . . . . . . | 5.22 |
| Lime . . . . . . | 1.64 |
| Oxide of nickel . . . . | 0.82 |
| " chromium . . . . | 0.70 |
| Sulphur . . . . . . | 4.24 |
| Water (?) . . . . . . | 6.50 |
| | 100.44 |

Fischer and Duflos have analyzed some meteoric iron which fell at Brannau, on the 14th July, 1847. According to them it was thoroughly soluble only in aqua regia, and consisted of:—

| | |
|---|---|
| Iron . . . . . . | 91.882 |
| Nickel . . . . . . | 5.517 |
| Cobalt . . . . . . | 0.529 |
| Residue . . . . . . | 2.072 |
| | 100.000 |

This residue was composed of copper, manganese, arsenic, calcium, magnesium, silicon, carbon, chlorine, and sulphur.*

As seen from the last analysis, meteoric iron is not always capable of being dissolved in hydrochloric acid, and it is therefore necessary to experiment on its solubility before regularly entering upon the analysis.

The rules given for crude iron apply in every respect to this occurrence of the metal, and it would be a repetition to detail them here.

### LXVI.—SILICATE OF PROTOXIDE OF IRON—FURNACE SLAG.

The carefully pulverized substance should be placed in an evaporating dish, thoroughly mixed with concentrated hydrochloric acid, some nitric

* Rammelsberg's *Repertorium d. Chem. Theils der Mineralogie.*

acid added to elevate the stage of oxidation of the iron, and the whole then digested until it becomes a homogeneous, yellow, gelatinous mass. This should then be evaporated to dryness in a sand-bath.

The residue should be digested in acidulated (muriatic) water, the silica removed by filtering, and then washed, dried, and heated to redness.

From the filtrate, the peroxide of iron is precipitated with ammonia, washed, dried, heated to redness, and weighed. From its weight the protoxide of iron is calculated.

If a slag contain lime and magnesia, they should be extracted from the filtrate of the iron precipitate, the former by oxalic acid, the latter by phosphate of soda.

A small admixture of copper we are able to detect in the original solution by hydrosulphuric acid, and traces of phosphoric acid, oxides of chromium and vanadium, by heating a large quantity of the slag to redness with nitre, when we examine farther, as will be shown in Art. XCI.

### LXVII.—COPPER SLAG, SILICATES OF POTASH, LIME, MAGNESIA, LEAD, AND COPPER.

After carefully pulverizing the slag, we mix it with hydrochloric acid, evaporate to dryness, macerate in hot acidulated (muriatic) water, filter off the silica, and wash with hot water.

From the filtrate, lead and copper are thrown down by sulphuretted hydrogen, the precipitate then separated from the liquid portion by means of filtering, and washed with hydrosulphuric acid water, when copper and lead are treated and isolated, as in Art. XXXIX.

The remaining solution we concentrate by evaporation, and separate the other bases as in Arts. III. and VI.

### LXVIII.—NATROLITH OR MESOTYPE. SILICATES OF SODA AND ALUMINA WITH WATER.

By exposing a weighed quantity of the mineral, dried at 212°, to a red heat, the water is driven off and its amount ascertained by weighing again and deducting from the original weight.

For the rest of the analysis we must employ another portion which has not been calcined. It is necessary to pulverize it with the utmost care;

after which, placing it in a porcelain vessel, we pour some moderately strong hydrochloric acid over it, cover the vessel, and digest until the whole has been converted into a transparent, gelatinous mass.

To make the silicic acid insoluble, it is necessary then to evaporate to dryness, stirring all the while.

The remaining saline mixture should be dissolved in a little water (acidulated with a few drops of hydrochloric acid), and the solution then filtered off from the silica, which should be washed, dried, heated to redness, and weighed.

A slight excess of caustic ammonia will precipitate the alumina from the filtrate. After applying a little warmth we filter, wash, dry, heat to redness, and weigh.

If the mineral contained iron, as is the case with the yellow natrolith, this should be separated, as in Art. LXX., from the alumina.

The last filtrate, after concentrating, we evaporate to dryness in a weighed platinum crucible, and carefully increase the heat until the sal-ammoniac is driven off, when a red heat should be applied. The residue is chloride of sodium, which contains 0.53010 of soda, or 0.39318 of sodium.

A specimen from Faroe, according to Smithson, contained:—

| | |
|---|---|
| Soda . . . . . . | 17.0 |
| Silica . . . . . . | 49.0 |
| Alumina . . . . . . | 27.0 |
| Water . . . . . . | 7.0 |
| | 100.0 |

While another variety containing iron, examined by Thompson, and coming from Antrim, in the province of Ulster, in Ireland, consisted of:—

| | |
|---|---|
| Soda . . . . . . | 14.93 |
| Silica . . . . . . | 47.56 |
| Alumina . . . . . . | 26.42 |
| Oxide of iron . . . . . | 0.58 |
| Water . . . . . . | 10.44 |
| | 99.93* |

LXIX.—FELSPAR.

The variety called Adular, according to Berthier, is composed of:—

| | |
|---|---|
| Silicic acid . . . . . | 64.20 |
| Alumina . . . . . | 18.40 |
| Potash . . . . . | 16.95 |
| | 99.55 |

* Alger's Phillips's *Min.*

Most varieties contain some iron. Of these, Vauquelin has examined the green felspar from Siberia, and Professor Gustavus Rose the flesh-colored mineral from Lomnitz, in Austrian Silesia. The first of the following analyses is of the former, the second of the latter:—*

| | | |
|---|---|---|
| Silica . . . | 62.83 | 66.75 |
| Alumina . . . | 17.02 | 17.50 |
| Potash . . . | 13.00 | 12.00 |
| Oxide of iron . . | 1.00 | 1.75 |
| Lime . . . | 3.00 | 1.25 |
| | 96.85 | 98.25 |

The mineral should be ground up to a very fine powder, and then different methods may be adopted in its analysis.

One mode of procedure is to mix the pulverized felspar with about four or five times as much carbonate of baryta, and to expose it in a platinum crucible to a white heat, until it is on the point of melting.

Another plan is to mix the mineral with four times its weight of hydrate of baryta, from which we remove the water, and then to expose it in a silver crucible to an intense red heat until it melts.

* Rammelsberg's *Handwörterbuch d. Chem. Min.*

Water is then poured over the molten mass, and by degrees concentrated muriatic acid added in drops, until, with the aid of slight warmth, it has all been dissolved, with the exception of gelatinous silica. To make the latter insoluble, the whole solution should be entirely evaporated to dryness.

The saline mass is then to be digested with water containing a little muriatic acid, until the soluble portion dissolves, while the silica remains. We then filter and wash, dry, heat to redness, and weigh the latter.

The main part of the baryta we precipitate from the filtered solution by gradual and careful addition of diluted sulphuric acid, and the remainder, together with the alumina, by a mixture of carbonate of ammonia with caustic ammonia, added in slight excess. After allowing it to settle for the space of twenty-four hours, we filter and wash. The alumina is then to be extracted with hot, diluted sulphuric acid, and again precipitated by caustic ammonia. After filtering, washing, and heating to redness, we weigh.

The filtrate from the united precipitate of alumina and baryta contains the potash. To obtain this we concentrate the solution by evaporation, saturate with hydrochloric acid, evaporate to dryness, and then, in a covered crucible, heat the residue almost to redness, when chloride of potash is left, which contains 0.63182 of potash.

If the mineral contain iron, that metal is separated as in former analyses, and if soda, this is treated as in Art. I.

A different way of examining the felspar is the following.

We place it in a platinum crucible, and pour fluoric acid, of almost fuming strength, over it, and digest until perfect decomposition has taken place; or else, spreading the mineral powder in a flat platinum dish or basin, we moisten with water, and in a suitable, closing vessel of lead, expose it for some time to the action of the vapors of fluoric acid, procured by pouring concentrated sulphuric acid over some pulverized fluorspar strewed over the bottom of the leaden vessel, and then applying slight heat.

In either case it is necessary afterwards, at a moderately increased temperature, to evaporate to dryness. The residue is then composed of silico-fluorate of potash, and silico-fluorate of alumina. To expel the silico-fluoric acid, concentrated sulphuric acid is poured upon it, and slight warmth applied until all is driven off.

Sulphates of alumina and potash remain, and these we dissolve in very little water, and precipitate the alumina with caustic ammonia; which, however, as in the first method given, contains sulphuric acid, to remove which latter we dissolve the precipitate in hydrochloric acid, after first washing, and again add ammonia.

The filtrate from the first precipitation of the alumina we evaporate to dryness and heat with great care (so that no spattering ensues, which would occasion loss) until the ammoniacal salt is driven off. The residue of acid sulphate of potash we next heat to redness in an atmosphere of carbonate of ammonia, to neutralize it. This is done by holding a piece of this salt in the redhot crucible.

If soda also be present in conjunction with the potash as a sulphate, it is necessary to add hydrochloric acid to prepare the salt for the treatment with chloride of platinum. (See Art. I.)

In this mode of analyzing we determine the quantity of silicic acid directly, but it is necessary to calculate its amount from the loss, or else to ascertain it by a separate treatment with carbonate of soda.

LXX.—GLASS, SILICATES OF LIME, POTASH, AND SODA, SOMETIMES CONTAINING PROTOXIDE OF LEAD, &c.

It is necessary to make two analyses, one by melting with carbonate of soda to decide upon the amount of silica, the second by treating with fluoric acid to determine its alkaline contents.

The composition of glass varies considerably. The recipe given for *flint glass* is:—*

| | |
|---|---|
| Ground flint . . . . | 100 parts |
| Oxide of lead . . . | 30 " |
| Saltpetre . . . . . | 34 " |

or,

| | |
|---|---|
| Clean, white, well-washed sand | 60 parts |
| Oxide of lead . . . | 35 " |
| Potash . . . . . . | 30 " |
| Saltpetre . . . . . | 2 " |

while the common *window glass* consists of:—

| | |
|---|---|
| Sand . . . . . . | 100 parts |
| Potash . . . . . . | 25 " |

* Quarizius, *Anorg. Tech. Chemie.*

| | |
|---|---|
| Glaubersalt, decrepitated by exposure to the action of atmospheric air . . . | 8 parts |
| Charcoal . . . . | 2 " |
| Wood-ashes, sifted . . . | 170 " |
| Broken glass . . . . | 120 " |

*Green bottle-glass* is generally composed of:—

| | |
|---|---|
| Sand . . . . . | 130 parts |
| Ashes, from which the lye has been extracted . . . | 80 " |
| Table salt . . . . | 16 " |
| Potash . . . . . | 20 " |
| White arsenic . . . | 1 " |

For the first analysis, one part of the well-ground glass is mixed with three parts by weight of a flux, consisting of four parts of carbonate of potash and two of carbonate of soda, and thus exposed to a melting heat, after which we dissolve in hydrochloric acid, and evaporate to dryness. The soluble part is then to be dissolved in acidulated (hydrochloric) water, and filtered off. Silica remains, and should be washed on the filter.

From the filtrate the casual adulterations, such as iron, manganese, and alumina, which are even common ingredients of colorless glass, we precipitate with ammonia, after first adding some chlorine water to convert the protoxide of manganese into the peroxide.

After this, oxalic acid will throw down the lime as oxalate, which we heat to redness to convert it into the carbonate.

If protoxide of lead be contained in the glass, we should extract this first, by introducing hydrosulphuric acid gas into the filtrate from the silicic acid.

Arsenic, if present, will thus also be precipitated together with the lead. To separate them, we dissolve them both in hydrochloric acid, and then throw down the lead with sulphuric acid, or extract the arsenic as ammoniaco-arseniate of magnesia.

To ascertain the quantity of the alkalies, we treat a second portion of thoroughly ground glass as in the felspar analysis, either with fluoric acid, or by heating to redness with carbonate of baryta, and then proceed as given in the last chapter. We are then also capable of again examining the other bases.

## LXXI.—AUGITE (PYROXENE) AND HORNBLENDE. SILICATES OF LIME, MAGNESIA, IRON, MANGANESE AND ALUMINA.

A green variety of this mineral from Dalecarlia,

in Sweden, according to Rose, gave the first of the following analyses, while a specimen from Finland examined by Berzelius gave the second.

| | | |
|---|---|---|
| Silica . . . | 54.08 | 50.99 |
| Lime . . . | 23.47 | 20.00 |
| Protoxide of iron . | 10.02 | 21.00 |
| Magnesia . . . | 11.49 | 4.50 |
| Oxide of manganese . | 0.61 | 3.00 |
| | 99.67 | 98.50 |

Another green variety, analyzed by Rose, contained also 0.14 of alumina.

After pulverizing the mineral to the utmost extent of which we are capable, we melt it with four times its amount of carbonate of soda and potash.

This mixture we dissolve in hydrochloric acid, with the addition of some few drops of nitric acid, when the silica is rendered insoluble and farther treated as in Art. LXVIII.

As ammonia would precipitate besides the peroxide of iron also, if present, alumina and magnesia, it is necessary, in such cases, to proceed as follows:—

The solution, diluted by the washings, we gradually neutralize, while constantly stirring, by carbonate of soda, so that it is saturated with carbonic acid, and by the addition of the last amount of carbonate of soda the bicarbonate of this alkali is formed.

By this process, oxide of iron and alumina alone are thrown down, while lime and magnesia, together with the protoxide of manganese, are retained in solution.

Alumina and iron are then separated, after filtering, as in Art. XIX.

The filtrate from this precipitate is mixed, according to the quantity of manganese, with more or less hypochlorite of soda, and set aside for twenty-four hours in a closed vessel, by which the manganese is thrown down in the shape of a hydrate of the hyperoxide, which by heating to redness is convertible into the sesquioxide, one-fourth of its oxygen being evolved.

The solution, filtered off from this, is to be concentrated by boiling in a slanting retort, when it should be saturated by hydrochloric acid—care being taken to avoid spattering—and an excess of ammonia added, after which oxalic acid will precipitate the lime. Slight digestive heat should be applied, and a little time given for the precipitate to collect, after which it is filtered off and treated as in Art. VI.

The magnesia we precipitate as in the same chapter, in the shape of ammoniaco-phosphate of magnesia.

### LXXII.—BERYL. EMERALD.

Klaproth gives the contents of a beryl from Siberia, as:—

| | |
|---|---|
| Silica . . . . . . | 66.45 |
| Alumina . . . . . | 16.75 |
| Glucina . . . . . | 15.50 |
| Perox. of iron . . . . | 0.60 |
| | 99.30 |

A variety from Broddbo, according to Berzelius, also contained $\frac{9}{125}$ per cent. of oxide of tantalum, while another from Fossum, according to Sheerer, contained $\frac{9}{5000}$ per cent. of lime.

A specimen of emerald, analyzed by Klaproth, consisted of:—

| | |
|---|---|
| Silica . . . . . . | 68.50 |
| Alumina . . . . . | 15.75 |
| Glucina . . . . . | 12.50 |
| Peroxide of iron . . . | 1.00 |
| Lime . . . . . . | 0.25 |
| Oxide of chromium . . . | 0.30 |
| Water and loss . . . | 1.70 |
| | 100.00 |

According to Rammelsberg, the calculated composition, omitting the casual impurities, is:—

| | | | | |
|---|---|---|---|---|
| Silica . | 9 atoms | = | 5195.79 | = 69.80 |
| Alumina | 2 " | = | 1284.66 | = 17.26 |
| Glucina | 18 " | = | 962.52 | = 12.94 |
| | | | 7442.97 | 100.00 |

The mineral, after being properly prepared in the mortar, is to be melted in a platinum crucible with four times its weight of dry carbonate of soda. After this, it should be digested with an excess of muriatic acid until entire decomposition has taken place, when we evaporate to dryness, to render the silica insoluble. The residue is then to be digested in water slightly acidulated with hydrochloric acid, the solution filtered off from the silica and very gradually, while stirring, poured into a hot solution of carbonate of ammonia in excess. Alumina is thus precipitated and separated from the glucina, which dissolves. After digesting the precipitate for some time with the solution, the latter is filtered off and boiled until all the ammoniacal salt is discharged, by which the glucina is thrown down.

A different method is to mix the solution, filtered off from the silica, with concentrated caustic potash in excess, in which, on application of heat, both earths are entirely soluble, and then to dilute with water and maintain it for some time at a seething temperature, by which the glucina is precipitated, the alumina remaining in solution. This we precipitate with ammonia, after slightly acidulating the solution with hydrochloric acid, and heating with chlorate of potash, to remove the organic matter occasioned by the dissolving of minute particles of the filter.

To remove the large quantity of chloride of sodium we may also precipitate both earths with caustic ammonia, and only after washing treat them with carbonate of ammonia or caustic potash.

LXXIII.—CLAY, ARGILLITE. SILICA, ALUMINA, AND WATER, FREQUENTLY WITH ADMIXTURES OF POTASH, MAGNESIA, OXIDES OF IRON AND MANGANESE, FELSPAR, SAND, AND OTHERS.

The amount of water we ascertain by drying the clay at 212° Fah., heating to redness, and weighing again. Organic matter, if present, is also thus discharged in vapor.

The clay should be heated with concentrated sulphuric acid, most of the acid driven off by evaporation, and then hydrochloric acid added, and heat applied, to dissolve all except the silica, which is removed by filtration.

If the clay contains sand or felspar, the silica should be dissolved in a boiling, concentrated solution of carbonate of soda, while the sand and felspar remain.

The muriatic solution should be considerably diluted, and gradually saturated with carbonate of soda, as in Art. LXX., so that peroxide of iron and alumina are precipitated, while the bicarbonates of lime, magnesia, and manganese remain in solution. The separation of peroxide of iron from alumina, has been given in Art. XIX., and that of the other bases from one another in Art. XVIII.

We may also analyze the clay by melting it with three times its weight of an anhydrous mixture of four parts of carbonate of potash, and three of carbonate of soda, and then dissolving the whole in diluted hydrochloric acid, evaporating to dryness, dissolving again in hydrochloric acid, and filtering off the solution from the silica.

The separation of the various contents of the solution is performed as above.

For determining the amount of alkali, a separate quantity of the clay is melted with hydrate or carbonate of baryta, and treated as in Art.

LXVIII., the baryta and other bases being here also removed from the solution, by precipitating with caustic ammonia and carbonate of ammonia. After slightly warming the precipitate, the solution is filtered off, evaporated, and the residue heated to redness, when chloride of potassium or of sodium remains.

LXXIV.—COMMON LIMESTONE, CEMENT, OR MORTAR, AND MARL; CONSISTING OF CARBONATES OF LIME, MAGNESIA, IRON, MANGANESE, WITH ALKALINE CLAY AND PHOSPHATE OF LIME.

For the detection of *alkali*, large pieces of the substance, if carbonate of lime predominates, are heated to an intense white heat for half an hour, by which the clay contained in it, and in which the alkali occurs, is decomposed.

After this, we reduce the compound to a fine powder, and macerate in water. To the solution, we add a little carbonate of ammonia, evaporate, and filter off the calcareous precipitate; then saturate with hydrochloric acid, evaporate to dryness, and heat the remaining chloride of potassium or of sodium to a moderate red heat. If both are present, they should be separated as in Art. I., by chloride of platinum.

The quantity of *water of crystallization* is determined by heating to redness a portion of the substance (previously dried at 212°, to drive off the atmospheric moisture—for it is very hygroscopic—and weighed immediately afterwards) in a glass tube. The water is then either ascertained by the loss, or by weighing in a chloride of calcium tube, attached to the other end, and the original weight of which is known.

To ascertain the *carbonic acid*, two methods exist. The most simple one is to place a new portion of the compound in a weighed vial, and to pour nitric acid, of a known quantity, over it. The loss is then carbonic acid.

A second plan is to melt a weighed portion of the lime, or marl, with exactly four times its weight of borax-glass (*i. e.* borax previously molten to form a slag, or glass) in a platinum crucible. In this manner, all the carbonic acid is driven off, but also all the water, and it is, therefore, necessary to deduct the previously ascertained weight of this from the total amount of the loss.

To separate the *clay*, another well-ground portion of the substance is employed. This should be digested with very dilute nitric acid, and thus the carbonates and the phosphate of lime dissolved, while the clay is left, which is to be filtered off, heated to redness, and weighed. We may then farther investigate its properties as directed in the last article.

The treatment of the solution, and the isolation of its ingredients is effected as in Art. XXIII.

### LXXV.—FLUOR-SPAR.

This mineral is simply fluoride of calcium, with occasional and varying adulterations of iron, alumina, &c. Wenzel, in his chemical investigations of fluor-spar, Dresden, 1783, gives as its composition,

| | |
|---|---|
| Fluoric acid . . . | 32.50 |
| Lime . . . . | 56.67 |
| Iron and alumina . . | 10.83 |
| | 100.00 |

While Richter found it to contain

| | |
|---|---|
| Fluoric acid . . . | 34.85 |
| Lime . . . . | 65.15 |
| | 100.00 |

To determine the amount of fluorine, we may either do so indirectly, by first ascertaining the quantity of lime, and then calculating how much of the fluorine is necessary to convert that into fluoride of calcium; or directly, by the loss occasioned when the mineral is dissolved in sulphuric acid, as in this manner all the fluoric acid is evolved; or as directed in the following article.

For the first process, we dissolve in sulphuric acid, the only acid in which the mineral is properly soluble, when fluoric acid is discharged in its gaseous form. From the solution, oxalic acid will precipitate the lime alone in the shape of the oxalate, convertible into the carbonate, by heating to redness. This contains 56 per cent. of calcium, and it is then necessary to calculate the amount of fluorine required to transform the lime into fluoride of calcium, which is composed of

| | |
|---|---|
| Calcium . . | 51.48 |
| Fluorine . . | 48.52 |
| | 100.00 |

The filtrate produced after precipitating the lime, contains all the other ingredients of the mineral, such as iron and alumina, and these are separated as in Art. XIX.

### LXXVI.—TOPAZ.

$$2\,(Al_2F_3) + 5\,(Al_2O_3,\ SiO_3).$$

Von Kobel has given the above formula for the pure composition of this mineral, though it also frequently contains peroxide of iron. The amount of this, however, varies very materially. According to Berzelius, a yellow specimen from the Brazils, contained,

| | |
|---|---|
| Silicic acid . . | 34.01 |
| Alumina . . . | 58.38 |
| Fluoric acid . . | 7.79 |
| | 100.18 |

On exposure to an intense white heat, the topaz loses all its fluorine in the shape of terfluoride of silicon ($3\,HO + SiF_3$).

By melting the carefully pulverized mineral with four times its weight of carbonate of soda, it is prepared for further treatment, and at the same time fluoride of sodium formed, which water will extract. It is, however, necessary, before filtering —and thus removing the silicate of alumina, which may also contain iron—to add a little carbonate of ammonia to the solution, and digest, and thus to precipitate a small portion of soluble alumina and silicic acid.

The residue is then to be removed by filtration, thoroughly washed, decomposed by hydrochloric acid, and treated as has been directed in the analysis of natrolite or mesotype.

The alkaline filtrate should be concentrated by evaporation, and then, being placed in a platinum dish, mixed with a small excess of hydrochloric acid, and set aside in some warm spot for twenty-four hours, to expel the carbonic acid from the carbonate of soda. After that, an excess of pure ammonia is added, the whole poured into a vial, which admits of being closed, and in this, mixed with a solution of chloride of calcium. All the fluorine is thus precipitated as fluoride of calcium. After giving sufficient time for this to collect and settle in the closed vial, we should decant the liquid portion, and add pure water, which is again to be poured off, and then filter, and wash, dry, and heat the precipitate to redness.

The result thus produced is only an approximate one, as some of the fluorine is retained by the alumina.

### LXXVII.—DATOLITE.

Professor Rammelsberg has analyzed this mineral from Arendal, in Sweden, and from Andreasberg, on the Hartz Mountains, in Germany, and ascertained its contents to be the following :—

| | ARENDAL. | ANDREASBERG. |
|---|---|---|
| Silica . . | 37.520 | 38.477 |
| Lime . . | 35.398 | 35.640 |
| Boracic acid | 21.377 | 20.315 |
| Water . . | 5.705 | 5.568 |
| | 100.000 | 100.000 |

From the mean of these, and other analyses, he gives as its composition :—

| | | | | | |
|---|---|---|---|---|---|
| Silica . | 4 atoms | = | 2309.24 | = | 37.910 |
| Lime . | 6 " | = | 2136.12 | = | 35.068 |
| Boracic acid | 3 " | = | 1308.60 | = | 21.482 |
| Water . | 3 " | = | 337.44 | = | 5.540 |
| | | | 6091.40 | | 100.000 |

A variety of the datolite, termed Botryolite, occurring in the Kjenlie mine near Arendal, which differs only in possessing twice as much water, according to the same authority, contains:—

| | |
|---|---|
| Silica . . . | 36.085 |
| Lime . . . | 35.215 |
| Boracic acid . . | 19.340 |
| Water . . . | 8.635 |
| | 99.275 |

The calculated composition therefore is:—

| | | | | |
|---|---|---|---|---|
| Silica . | 4 atoms | = 2309.24 | = | 35.920 |
| Lime . | 6 " | = 2136.12 | = | 33.227 |
| Boracic acid | 3 " | = 1308.60 | = | 20.355 |
| Water . | 6 " | = 674.88 | = | 10.498 |
| | | 6428.84 | | 100.000* |

To ascertain the amount of water, a weighed portion of the mineral should be exposed to an intense red heat.

For the farther analysis, a different portion is to be thoroughly reduced to powder, digested with hydrochloric acid, and afterwards the temperature increased nearly to the boiling point, so that on cooling the whole is converted into a gelatinous mass. The silica, after first having been rendered insoluble, is then removed by filtration, washed with hot water, heated to redness and weighed.

Ammonia in excess is next added to the solu-

* Rammelsberg, *Handwörterbuch. d. Chem. Min.* vol. i. pp. 186–188.

tion, and the lime, together with a little silica, precipitated either with oxalic acid or carbonate of ammonia.

The solution, filtered off from this, should be evaporated to dryness, while during that process a little ammonia is repeatedly added, to avoid the evaporation of boracic acid. When the saline mass has become perfectly dry, it should be carefully heated until the ammoniacal salts are driven off, when pure boracic acid remains, the weight of which, however, is always somewhat deficient.

By dissolving the boracic acid in water, a slight admixture of silica may be extracted.

The quantity of boracic acid is more accurately to be ascertained by the loss; and to do this, we first, as above, extract the silica, then neutralize the filtrate with ammonia, and precipitate with carbonate of ammonia, when, as above, the lime is procured. The loss is boracic acid and water.

## LXXVIII.—ZIRCON AND HYACINTH; CHIEFLY SILICATE OF ZIRCONIA.

Professor Rammelsberg, of Berlin, gives the following as the calculated composition of zircon and hyacinth, deduced from various analyses, some of which are given below.

| | | | | | |
|---|---|---|---|---|---|
| Silica . | 1 atom | = | 577.31 | = | 33.61 |
| Zirconia . | 1 " | = | 1140.40 | = | 66.39 |
| | | | 1717.71 | | 100.00 |

| | FROM CEYLON. | | EXPAILLY, IN AUVERGNE. |
|---|---|---|---|
| | Zircon. | Hyacinth. | Hyacinth. |
| Silica . | 26.5 | 25.0 | 33.3 |
| Zirconia . | 69.0 | 70.0 | 66.7 |
| Perox. of iron | 0.5 | 0.5 | — |
| | 96.0 | 95.5 | 100.0 |
| | KLAPROTH. | | BERZELIUS. |

To Klaproth, formerly a professor of chemistry at Berlin, the merit is due of having discovered Zirconia, A. D. 1789.

For the analysis, we should carefully select the purest pieces of crystals, and heat them to redness, by which they lose their color. After pulverizing, the finest powder is washed out from the coarser particles, and melted with four times its weight of carbonate of soda, in a good fire. The mass is then digested with water, which dissolves the silicate of soda formed, and leaves as a crystalline powder, a combination of zirconia and soda. After filtering and washing, the residue is to be covered with concentrated hydrochloric acid, and the salt dissolved in water, when the zirconia should be precipitated by ammonia.

According to Vanuxem, the zircon of North Carolina contains:—

| | |
|---|---|
| Silica . . . . . . | 32.08 |
| Zirconia . . . . . . | 67.07 |
| | 99.15 |

### LXXIX.—THORITE AND THORINA.

Berzelius analyzed this mineral from Loev-oen, near Brevig, in Norway, and in it discovered a new earth, thorina, which has not been found to exist in any other compound. His analysis gave, as the composition of thorite:—

| | |
|---|---|
| Silica . . . . . . | 18.98 |
| Thorina . . . . . . | 57.91 |
| Lime . . . . . . | 2.58 |
| Peroxide of iron . . . | 3.40 |
| Peroxide of manganese . . | 2.39 |
| Magnesia . . . . . | 0.36 |
| Oxide of uranium . . . | 1.61 |
| Peroxide of lead . . . | 0.80 |
| Peroxide of tin . . . | 0.01 |
| Potash . . . . . . | 0.14 |
| Soda . . . . . . | 0.10 |
| Alumina . . . . . . | 0.06 |
| Water . . . . . . | 9.50 |
| Insoluble powder . . . | 1.70 |
| | 99.51 |

Rammelsberg quotes the process adopted by Berzelius in the analysis of this complicated mineral,* the translation of which is given here.

"A part of the mineral, in the shape of a coarse powder, was heated to redness in a little retort, and the water thereby discharged collected in a chloride of calcium tube. The loss showed that some other ingredient had been discharged as vapor.

"The well-pulverized mineral (not submitted to a red heat) was sufficiently moistened with hydrochloric acid, when some chlorine was evolved. On applying heat the mass became gelatinous. It was then evaporated to dryness. On redissolving it, the silica was left insoluble, and by boiling this with a solution of carbonate of soda it was separated from some undecomposed mineral and quartz.

"The filtrate from the silica residue was precipitated with caustic ammonia, and the filtrate from this again by oxalic acid, and thus the lime thrown down. By heating to redness this was converted into the carbonate, and then freed from a little peroxide of manganese by dissolving in hydrochloric acid, and adding first some bromine water and then ammonia.

---

* Ram. *Handwörterbuch. d. Chem. Min.* vol. ii. p. 213.

"The solution, after removing the oxalic precipitate, was evaporated to dryness, the sal-ammoniac driven off, and the residue treated with water, which left insoluble magnesia.

"The aqueous solution, on being evaporated, gave a mixture of chlorides of potassium and sodium, which were separated by chloride of platinum.

"The precipitate, produced above by ammonia, was dissolved in hydrochloric acid, when hydrosulphuric acid gas gave a black precipitate in the diluted liquid. This was oxidized with nitric acid, and then evaporated with some sulphuric acid, until the surplus acid was driven off. On treating the residue with water a solution was obtained, from which ammonia precipitated peroxide of tin, while the remainder consisted of sulphate of lead.

"The solution, from which the lead and tin were removed by sulphuretted hydrogen, was reduced to dryness, and water allowed to act upon the residue, when some silica remained insoluble. The solution was boiled with an excess of caustic potash, by which a small proportion of alumina was found to be dissolved.

"The portion precipitated by the potash, on being dissolved in hydrochloric acid, left some peroxide of manganese, which contained traces of peroxide of iron and alumina.

"The muriatic solution was neutralized with ammonia, concentrated by evaporation, and mixed with sulphate of potash, which was added until enough was dissolved to saturate the solution. A precipitate of potassio-sulphate of thorina was produced, which was washed with a solution of sulphate of potash, dissolved in almost boiling water and reprecipitated with potash. The precipitate was the thorina, containing a trace of manganese.

"The solution, from which the potassio-sulphate of thorina had been removed, was precipitated by potash and the precipitate treated with carbonate of ammonia. The insoluble parts were a mixture of the peroxides of iron and manganese, which were separated in the common way by succinate of ammonia.

"The solution in carbonate of ammonia was evaporated to dryness, and the residue digested with dilute acetic acid, which extracted the oxide of uranium. This was precipitated by ammonia, and heated to redness.

"The portion left by the acetic acid was of a yellowish-brown color. With hydrochloric acid it gave a colorless solution, in which tartaric acid, ammonia, and hydrosulphate of ammonium showed the presence of iron, while the solution, after precipitating that and then evaporating, and heating the residue to redness, left some thorina."

From the formation and emission of chlorine gas on dissolving the mineral, we deduce that iron and manganese are present in the shape of peroxides.

Berzelius considers the thorite as a mixture of hydrated silicates, in which the chief ingredient is a silicate of thorina, consisting of one atom of silicic acid to three of thorina, and of which this mineral contains 71.5 per cent.

## LXXX.—TRIPHYLINE, TETRAPHYLINE,* PETALITE, AND SPODUMENE.

The following analyses of the Triphyline and Tetraphyline are given by Rammelsberg, the first from Bodenmais by Fuchs, the second by Berzelius and Nordenscioeld, of a specimen of the latter variety from Keiti, in Finland.

* Named by Nordenscioeld.

| | TRIPHYLINE. | | TETRAPHYLINE. |
|---|---|---|---|
| Phosphoric acid . | 41.47 | | 42.6 |
| Protoxide of iron . | 48.57 | | 38.6 |
| Protoxide of manganese | 4.70 | | 12.1 |
| Lithion . . . | 3.40 | | 8.2 |
| Silica . . . | 0.53 | Magnesia | 1.7 |
| Water . . . | 0.68 | | |
| | | | 103.2 |
| | 99.35 | | |

The thoroughly pulverized mineral should be dissolved in hydrochloric acid, the solution boiled with nitric acid, and all the oxide of iron and phosphoric acid precipitated with ammonia. After precipitating the manganese from the filtrate, with hydrosulphate of ammonia, the solution is evaporated to dryness, and the remaining salts heated until all the chloride of ammonium is driven off. The residue is chloride of lithium. (LCl) soluble in alcohol.

In analyses, as for example, of Petalite or Spodumene, where we wish to separate the chlorides of lithium and sodium, we macerate the dried combination of the two in an anhydrous mixture of alcohol and ether, which extracts the lithion salt.

## LXXXI. — CERITE AND OCHROITE; SILICATES OF CERIUM, LANTANIUM, AND DIDYMIUM.

The finely pulverized cerite is moistened considerably with concentrated sulphuric acid, and digested with it, some drops of nitric acid being added, to oxidize the very common admixture of sulphuret of bismuth.

The whole is then to be macerated in cold water, without applying heat during the process, until the sulphates are dissolved. After removing the silica from the solution, the latter, if too much diluted by the washings, should be concentrated and mixed with a seething-hot, concentrated solution of sulphate of potash, and with it left to cool. By this all the three bases are precipitated as bisulphates, while iron, &c., remains in solution. The precipitate is then to be separated from the liquid portion by filtration, and washed with a saturated solution of sulphate of potash. Into the filtrate, some crystals (crystalline crust, or sediment), of sulphate of potash are placed, and by that some remaining portion of the double salt is precipitated.

The bisulphates produced in this manner, should

be dissolved in seething water, hydrochloric acid being added. The bases are then to be precipitated from the hot solution by an excess of caustic potash.

After heating to redness, they form a cinnamon-colored powder. By digesting in concentrated sulphuric acid, and converting them into a salt, this gives a yellow solution with water, from which sulphate of potash will throw down a lemon-colored mixture of double salts.

If the brown oxide of cerium is macerated in water, to which, as it becomes saturated, we add nitric acid, drop by drop, the greater part of the oxide of lantanium is extracted. This is precipitated by carbonate of ammonia, and on heating to redness, becomes almost entirely white.

The manner of isolating these three metals is not yet perfectly known, and the didymium is at present not even fully described. In the analysis of the cerite they are, therefore, not all of them separated, as is seen from the following analysis by Hermann :—

| | |
|---|---|
| Carbonic acid . . | 4.62 |
| Silica . . . . | 16.06 |
| Protoxide of cerium . | 26.55 |
| Peroxide of lantanium . | 33.38 |
| Water . . . . | 9.10 |
| Alumina . . . | 1.68 |
| Peroxide of iron . . | 3.53 |
| Lime . . . . | 3.56 |
| Peroxide of manganese . | 0.27 |
| Copper . . . | Trace |
| | 98.75 |

It should be remarked that the specimen here examined was considerably adulterated by calcspar, from which the whole of the carbonic acid originated.* Klaproth has analyzed another occurrence of this mineral, for which Hermann proposes the name of ochroite. It contained :—

| | |
|---|---|
| Silica . . . . | 34.50 |
| Protoxide of cerium . | 50.75 |
| Peroxide of iron . . | 3.50 |
| Lime . . . . | 1.25 |
| Water . . . . | 5.00 |
| | 95.00 |

* Rammelsberg's *Handwtbch. d. Chem. Min.* supplement, ii. p. 32.

This mineral Berzelius considers as consisting, in its purest form, of one atom of silica, three atoms of protoxide of cerium, and three of water.*

LXXXII.—CHROMATED IRON.

Berthier has analyzed a specimen of this mineral from the Bare Hills, in Maryland, and gives his result as:—

| | |
|---|---|
| Oxide of chromium . . | 51.6 |
| Peroxide of iron . . | 35.0 |
| Alumina . . . . | 10.0 |
| Silica . . . . | 3.0 |
| | 99.6 |

1. For the qualitative investigations only, the well-pulverized mineral is mixed with an equal proportion of nitre, and as much carbonate of potash, and exposed for at least half an hour to an intense red heat. On cooling, the chromate of potash formed is extracted by water.

The residue, consisting of oxide of iron, alumina, and magnesia, is to be dissolved in hydrochloric acid, when generally some undecomposed mineral remains undissolved. The three oxides are separated as below.

* Rammelsberg's *Handwtbch. d. Chem. Min.* vol. i. 146.

The solution of chromate of potash commonly contains a little alumina, silica, and manganic acid; to precipitate which, it is necessary to mix it with carbonate of ammonia, and boil.

To produce bichromate of potash from the solution, this is mixed with an excess of nitric acid, concentrated by evaporation, and set aside for the salt to crystallize.

To precipitate chromate of lead, the solution is mixed with acetic acid in excess, and acetate of lead added.

To procure the chromium from its solution, in the shape of sesquioxide of chromium, the liquid ought to be more than saturated with sulphuric acid, when a sufficient quantity of sulphurous acid is added, to convert the color into an emerald green. The temperature of the solution is then elevated almost to boiling, and the sesquioxide of chromium precipitated by caustic ammonia, washed and heated to redness.

A second method, to produce this compound, is exactly to saturate the yellow solution with nitric acid, and then to precipitate the chromic acid with nitrate of mercury (protoxide). After washing on the filter, drying and heating to redness, the yellowish red chromate of mercury (protoxide) leaves pure green sesquioxide of chromium.

2. For the quantitative analysis, the mineral, exceedingly carefully pulverized, is melted with four times its weight of bisulphate of potash in a platinum crucible. It is necessary to take the greatest precaution that no portion of the mass is thrown out and lost by the ebullition, produced on first applying heat. By degrees the temperature is increased to a red heat, and the mixture maintained in a redhot fluid state for some time.

The salts formed are hardly soluble in water, and it is, therefore, necessary to convert the sesquioxide of chromium into a chromated alkali, which is effected by spreading about a double quantity (in volume) of a mixture of equal parts of nitre and carbonate of soda, over the mass, previously allowed to cool in the crucible, and to apply heat until a perfectly fluid state is produced.

After this again has become cold, the chromate of potash is extracted with water, and the residue of peroxide of iron, alumina, magnesia, and silica washed out completely, dissolved in concentrated hydrochloric acid, and treated as in Art. LXX.

The aqueous solution of chromate of potash should be mixed with an excess of hydrochloric acid, and brought to a seething heat, while, at the same time, alcohol is added, drop by drop, until the liquid assumes an emerald green color, when the sesquioxide of chromium is precipitated by caustic ammonia, and, after heating to redness, weighed as such. The sesquioxide of chromium contains 0.68645 parts of chromium, and one part of the oxide is equal to 1.31355 of chromic acid.

LXXXIII. — NATURAL CHROMATE OF LEAD, AND CHROME YELLOW OF COMMERCE; OR, CHROMATE OF LEAD, WITH FREQUENT ADMIXTURES OF WHITE CLAY, CARBONATE OF LIME, AND SULPHATES OF BARYTA, LIME, AND LEAD.

Berzelius has published the following analysis of pure natural chromate of lead, in Schweigger's *Journal of Natural Philosophy and Chemistry* (published in Nuremberg and Halle, Germany), vol. xxxiv. p. 72:—

| | |
|---|---|
| Oxide of lead . . . | 68.50 |
| Chromic acid . . . | 31.50 |
| | 100.00 |

showing that there is one atom of each, so that the calculated composition is,

| | |
|---|---|
| Oxide of lead . . . | 68.15 |
| Chromic acid . . . | 31.85 |
| | 100.00 |

It is hardly necessary, after stating this, to remark, that pure chromate of lead should give in its analysis almost exactly the same result as that deduced from the formula $PbO,CrO_3$.

To analyze it, the compound should be digested with a mixture of fuming hydrochloric acid and alcohol, by which a green solution of chlorate of chromium is produced, while the lead remains as an insoluble white powder of chloride of lead. The latter is to be collected on a filter, dried at 212°, and washed with alcohol. After diluting the filtrate with water and evaporating the alcohol, the sesquioxide of chromium is to be precipitated from the seething solution by ammonia, and, after heating to redness, weighed, as in the last article, q. v.

In analyzing a chrome yellow, adulterated by the substances mentioned in the heading of this article, it is necessary to boil it repeatedly with a large quantity of water, thus to extract the gypsum (sulphate of lime).

The washed residue is then to be digested in

very dilute nitric acid, until effervescence is no longer produced. By this the carbonate of lime is extracted.

After washing again, the residue should be digested in somewhat concentrated tartrate of ammonia, containing an excess of the alkali. The sulphate of lead is thus dissolved, and can be precipitated from its solution by sulphuret of potassium or chromate of potash.

A mixture of chromate of lead, sulphate of baryta, and clay remains. This is treated with hydrochloric acid and alcohol, and the chloride of lead, thus formed, separated from the baryta salt and clay by boiling with water.

The mixture of the two latter is heated with concentrated sulphuric acid until most of the surplus acid is driven off, when the sulphate of alumina is extracted by water, and the alumina precipitated from the filtrate by ammonia.

To extract the silica from the remaining mixture of that substance with sulphate of baryta, the whole is boiled with a concentrated solution of carbonate of soda, from which, after filtering, chloride of ammonium will precipitate the silica.

### LXXXIV.—TUNGSTATE OF IRON.

The composition of this mineral without the casual admixtures, is, according to Rammelsberg:—

| | |
|---|---|
| Tungstic acid . . . | 77.09 |
| Protoxide of iron . . | 17.11 |
| Protoxide of manganese . | 5.80 |
| | 100.00 |

Oxide of tin and silica sometimes occur in it as impurities.

1. *Qualitative Analysis and Preparation of Tungstic acid.*

The pulverized mineral is to be digested with a mixture of four parts of concentrated hydrochloric acid, with one of nitric, until it is converted into a yellow powder of tungstic acid. This is to be collected on a filter, washed, dissolved in caustic ammonia, and the solution filtered and allowed to crystallize by evaporating. After heating to redness with access of atmospheric air, the salt will leave pure yellow tungstic acid.

A second plan is to mix six parts by weight of the well-ground mineral with three of carbonate of potash and one of nitre, and thus to expose it to a red heat for the space of half an hour. After it has become cold, water will extract the tungstate of potash formed.

From this solution the tungstic acid may be procured in three different ways.

One method is to neutralize the solution with nitric acid, and add nitrate of mercury as long as a precipitate is produced. The salt thus thrown down is washed out on the filter, dried and placed in a glass retort, when the mercury is distilled off, while pure tungstic acid is left.

Another plan is to neutralize the solution with chloride of calcium, to wash the tungstate of lime, thus precipitated, and then to boil it in hydrochloric acid, when, after filtering and drying, pure yellow tungstic acid remains.

A third method is to add to the solution twice as much powered sal-ammoniac as the carbonate of potash previously used, to evaporate to dryness, place the mixture in a crucible, cover it with a layer of table-salt, and to apply heat until no more vapors of sal-ammoniac are emitted, when the temperature is increased to redness. On macerating the mass in water, black crystalline oxide of tungsten remains, which should be leached, and by heating to redness, while exposed to the action of the atmospheric air, converted into tungstic acid.

2. *Quantitative Analysis.*

In the quantitative analysis of the tungstate of iron, we digest the washed mineral powder in a mixture of four parts of concentrated hydrochloric acid, and one of nitric acid, until it has become entirely decomposed, when the solution is evaporated—towards the end of the process in a water-bath—to dryness, and the chloride of manganese and chlorate of iron dissolved and filtered off. Tungstic acid remains, and is washed out with alcohol, heated to redness and weighed.

From the diluted (with water) solution of the basic oxides of the two metals, the alcohol is first removed; they are then separated, as in Art. XVIII. They generally also contain some lime.

If the tungstic acid contain silica, the amount of this may be ascertained by dissolving the former in ammonia.

A second method for the quantitative analysis is the following:—

Six parts of the washed mineral are heated to redness in a platinum crucible with four of carbonate of potash and one of saltpetre. The mixture is then macerated in water, and the remaining oxides are carefully washed; after which the solution is to be neutralized with nitric acid, and the tungstic acid precipitated from it by nitrate of mercury, to effect which it is necessary, towards the end of the process, to neutralize the surplus of nitric acid by a few drops of ammonia. A black precipitate is produced, which ought to be carefully washed, and after drying heated to redness, when pure tungstic acid remains. This acid consists in a hundred parts of

| | |
|---|---|
| Tungsten . . . . . | 79.77 |
| Oxygen - . . . . . | 20.23 |
| | 100.00 |

The mixture of the oxides of iron and manganese is dissolved in concentrated hydrochloric acid, when generally a small portion of tungstic and silicic acids are left. The two oxides are then separated from one another as in Art. XVIII.

### LXXXV.—TUNGSTATE OF LEAD.

This mineral is of exceedingly rare occurrence. The only known locality is Zinnwald, in Bohemia. According to Lampadius, formerly professor at the Mining Academy of Freiberg, in Saxony, it contains:—

| | |
|---|---|
| Oxide of lead . . . | 48.25 |
| Tungstic acid . . . | 51.75 |
| | 100.00 |

This mineral occurs only in very minute crystals, which it is necessary to remove from the matrix with a pair of forceps, before reducing them to powder. We dissolve, as in the last article, in aqua regia, evaporate to dryness, redissolve the lead and filter. Sulphuric acid will precipitate the lead from the filtrate, while the tungstic acid which remains is washed with alcohol while on the filter, dried, heated to redness and weighed.

### LXXXVI.—TUNGSTATE OF LIME.

According to professor Rammelsberg, a fine yellowish red crystalline variety of this mineral, occurring near Hartzgerode, on the Hartz Mts. of Germany, was composed of

| | |
|---|---|
| Lime . . . . . | 21.56 |
| Tungstic acid . . . | 78.64 |
| | 100.20 |

Other varieties contain, besides these, some foreign admixtures, such as silica, and oxides of manganese and iron.

The analysis of this mineral is effected according

to the first method given for the quantitative analysis in Art. LXXXIV. The lime is precipitated with oxalic acid and then converted into the carbonate by heating to redness.

### LXXXVII.—IRON PYRITES.

This very common mineral consists in its pure composition, without bringing any of the casual admixtures into consideration, of

| | |
|---|---|
| Iron . . . . . . | 45.74 |
| Sulphur . . . . . | 54.26 |
| | 100.00 |

being a bisulphuret of iron. This mineral, however, rarely occurs in its perfectly unadulterated form. Arsenic, nickel, copper, and gold, are not uncommon admixtures.

The arsenic is very easily detected in the following manner. Small particles of the iron pyrites are placed in a little glass tube, closed at one end, and held in the flame of a spirit-lamp. The sulphur is the first substance distilled over, presenting at the beginning the appearance of a white amorphous powder, which gradually deepens into the lemon yellow of flowers of sulphur, and still later, if arsenic be present, assumes a tulip red, caused by the union of the two substances.

In the analysis, we first digest in nitric acid to convert the sulphur into sulphuric acid, dilute the solution, and decant it from the insoluble residue, which consists in part of gold, if this be present in the mineral.

The solution, besides iron, contains the copper, nickel, and arsenic, if the iron pyrites contained these. They are separated as in Art. LV.

### LXXXVIII.—BISULPHURET OF MOLYBDENUM.

The composition of this mineral, as deduced from the mean of the various analyses, is,

| | |
|---|---|
| Sulphur . . . . | 41.08 |
| Molybdenum . . . | 58.92 |
| | 100.00 |

In this calculation, the newer atomic weight of molybdenum, ascertained by Svanberg and Struve, has been employed. It is 575.83, while formerly the equivalent was considered to be 598.52.

The finely pulverized mineral is placed in an inclined crucible, and thus roasted at a moderate heat, while constantly stirring, until the odor of sulphurous acid is no longer perceptible.

From the roasted mineral, the molybdic acid is extracted by digesting with diluted ammonia. The solution, after being filtered off, is evaporated to dryness; the salt redissolved in diluted ammonia; the remaining impurities removed by filtration, and all the molybdic acid precipitated from the slightly ammoniacal solution, by bichloride of mercury. The washed and dried precipitate leaves, on heating to moderate redness with access of atmospheric air, or in a current of oxygen, pure molybdic acid, which contains 65.72 per cent. of molybdenum, and 34.28 of oxygen.

## LXXXIX.—ANALYSIS OF PITCH-BLENDE, OR UNCLEAVABLE URANIUM ORE, AND PREPARATION OF PURE PEROXIDE OF URANIUM.

According to professor Rammelsberg, a specimen of this mineral from Joachimsthal, in Bohemia, consisted of

| | |
|---|---|
| Perox. and protox. of uranium | 79.148 |
| Lead . . . . . . | 6.204 |
| Bismuth (cont. lead and copper) | 0.648 |
| Iron . . . . . . | 3.033 |
| Arsenic . . . . . | 1.126 |
| Lime . . . . . . | 2.808 |
| Magnesia . . . . . | 0.457 |
| Silica . . . . . . | 5.301 |
| Water . . . . . . | 0.362 |
| | 99.087 |

Some varieties of this mineral have also been found to contain vanadic acid and selenium; nor are nickel, cobalt, antimony, zinc, manganese, and sulphur quite uncommon admixtures.

After reducing the mineral to a very fine-grained powder, it is to be digested in moderately diluted sulphuric acid, to which, by degrees, some nitric acid is added. This process is continued until it is in part dissolved, and in part converted into a white powder. After this, the chief portion of the surplus sulphuric acid is driven off, the mass macerated in water, and digested. On cooling, and after the white precipitate has collected, the solution is to be filtered off.

The white residue consists of silica, sulphate of lead, and basic sulphate and arseniate of bismuth, which are separated as already directed.

The filtrate should be heated up to about 140° F., and while at this temperature, saturated for some time with sulphuretted hydrogen. While the current of this gas is yet passing through it, we allow the solution to cool down to the temperature of the atmosphere, and when satisfied as to its perfect saturation with the hydrosulphuric acid gas, cover it up and leave it thus for twenty-four hours. After the lapse of that time, the gas is driven off by the application of moderate warmth, and the precipitate removed by filtration.

This consists of arsenic, antimony, copper, and a small portion of lead and bismuth. The rules for the separation of them have been given on a previous page.

The solution, filtered off from these, should be brought to a seething temperature, and while in that state, gradually mixed with fuming nitric acid, until all the protoxide of iron is again converted into the peroxide, and the solution itself has once more assumed a clear yellow color. We then precipitate with an excess of caustic ammonia, and collect the yellowish-brown precipitate on a filter.

A part of the nickel, cobalt, zinc, lime, and magnesia (considering, of course, all these as having existed in the mineral), remain in solution, while the other portion is precipitated with the oxides of uranium and iron.

After having washed the precipitate, a hot and rather concentrated solution of carbonate of ammonia is poured over it, and heat applied until it has assumed the aspect of hydrated peroxide of iron. The uranium has then in part been dissolved, and its solution should be filtered off as quickly as possible, and while it is yet warm. The residue of uraniferous peroxide of iron should be thoroughly washed with water, and the washings collected separately.

The yellowish, or, when cobalt is present, reddish-yellow filtrate, if it be sufficiently concentrated, will, on being set aside for some time and allowed to cool, form a crystalline deposit of pure ammoniaco-carbonate of uranium, which ought to be collected and several times leached in cold water. On heating this to redness, pure dark-green peroxide and protoxide of uranium are produced.

The mother-lye, together with the washings, is carefully, and drop by drop only, mixed with hydrosulphate of ammonia, as long as a brownish-black precipitate is formed. We then filter immediately.

This precipitate consists of sulphurets of cobalt, nickel, and zinc, which are to be separated in the manner shown in a previous article.

The filtered yellow solution is to be heated to a seething point, until almost all the ammoniacal salt has been evolved, and all the oxide of uranium is thrown down.

The precipitate, which possesses a pure yellow color, and is uraniate of ammonia, should be removed by filtration. When it begins to pass through the filter with a milky appearance, it is necessary to wash with a solution of chloride of ammonium (sal-ammoniac).

By bringing this residue to a red heat, peroxide and protoxide of uranium are formed. By digesting it with diluted hydrochloric acid, the lime and magnesia may be extracted. They are then separated as in Art. VI.

To procure the oxide of uranium, chemically combined with the hydrated peroxide of iron, the latter should be dissolved in the smallest necessary quantity of hydrochloric acid, the solution neutralized with carbonate of ammonia, and, while stirring it, some carbonate and hydrosulphate of ammonia, in mixed solutions, added in drops. By this, all the iron is precipitated as protosulphuret of iron, while the oxide of uranium remains suspended in the solution. By boiling the filtered solution, the latter is thrown down.

Another plan to separate the two is, to reduce the peroxide of iron in an atmosphere of hydrogen, and then immediately to let the pyrophoric mass fall into diluted hydrochloric acid, which dissolves the iron, and leaves uranium as protoxide.

To discover the amount of arsenic, selenium, and

vanadium in pitch-blende, or other ores of uranium, the mineral should be mixed with twenty-five per cent. of a mixture of carbonate of soda and nitre, and thus heated to redness. Water will then extract seleniate, vanadiate, and arseniate of alkali. The separation of these is treated of in another article.

As seen in the analysis by professor Rammelsberg, it is not necessary to state the amount of uranium, otherwise than as a mixture of its two oxides, as it generally occurs in the mineral in this shape. By reducing in a current of hydrogen, protoxide of uranium is produced.

## XC.—TITANIATE OF IRON. SEPARATION OF TITANIC ACID.

Varieties of this mineral, named from their localities, are the iserine, from the stream Iser, in the Riesengebirge of Silesia), which, according to professor Henry Rose, consists of 50.12 per cent. of titanic acid, and 49.88 of protoxide of iron; the menaccanite from Menaccan, in Cornwall, and probably also the basonomelan of Von Kobell.*

Almost all varieties of titaniate of iron contain

* Alger's Phillips's *Mineralogy* (Boston, 1844, p. 378).

both peroxide and protoxide of iron, though the percentage of each differs considerably even in specimens from the same locality, and it is owing to this, that various theories are current as to the exact composition of the mineral. This is, however, not the place to enter upon their respective merits. Suffice it to say that, in giving the result of the analysis, it will be quite enough to ascertain the state of oxidation of the iron, and if both oxides occur, to separate them, as in Art. LXII., and to put down the titanium as titanic acid.

The finely pulverized mineral should be digested in concentrated hydrochloric acid, until all the iron is dissolved, and the titanic acid remains with a pure white color.

Or, otherwise, the mineral powder should be melted in a platinum crucible, with bisulphate of potash at a red heat, the mixed mass dissolved in water, the solution diluted, saturated with ammonia, and precipitated with hydrosulphate of ammonia, when a deposit of protosulphuret of iron and titanic acid is produced.

When the precipitate has settled, the supernatant liquid is to be decanted, and sulphurous acid poured over the former, by which all protosulphuret of iron is dissolved as dithionite of iron (syn. bisulphite of iron), while the titanic acid remains white and insoluble, and, after washing with dilute sulphurous acid and drying, should be heated to redness. To discharge an amount of sulphuric acid, a piece of carbonate of ammonia is held in the crucible.

A third plan is to heat the powdered mineral intensely in a current of hydrogen, and thus to reduce the iron which is easily extracted with hydrochloric acid.

### XCI.—IRON ORE CONTAINING VANADIUM.

To extract the vanadium, the properly ground ore is mixed with twenty-five per cent. of saltpetre and thus exposed to a moderate red heat for the space of half an hour. On cooling, the mass should be pulverized and boiled in water.

The filtered solution is of a yellow color, and contains vanadiate, chromate, phosphate, and nitrate of potash and alumina.

Nitric acid is added to it in slight excess, after which it should be boiled until the nitrous acid gas is driven off. While yet at a seething temperature ammonia is added, and thus the alumina precipitated as ammoniaco-phosphate of alumina.

The solution filtered off from this should be mixed with chloride of barium, which causes the precipitation of vanadiate, chromate, and phosphate of baryta. This precipitate is to be washed out on the filter, and while yet moist boiled in sulphuric acid until it is converted into sulphate of baryta. The filtrate from this contains the acids, and is of a yellowish-red color. It should be concentrated by evaporation and saturated by ammonia, when a piece of sal-ammoniac ought to be immersed in it. In the same measure as the solution becomes saturated by it vanadiate of ammonia is formed, and collects as a white or yellow crystalline powder. When the whole of it has been deposited we filter, and wash with a solution of chloride of ammonium. By heating with access of atmospheric air, vanadic acid, of a black color, with a shade of red, is left. A strong heat melts it to a red fluid, which, on cooling, stiffens into a crystalline mass.

From the solution which contains the chromic acid, this should be extracted, as directed in the article on chromated iron.

Another plan for the analysis of this iron ore is to treat the yellow solution (produced by macerating the mineral roasted with saltpetre) with sulphurous acid, until it has assumed a greenish blue color, then to neutralize with ammonia, add hydrosulphate of ammonia, and to heat to a seething point.

By this process, oxide of chromium and alumina are precipitated, while the vanadium remains in solution in the shape of a sulphuret, and can be precipitated from the filtered solution by diluted sulphuric acid. By melting it then with nitre, it is converted into vanadiate of potash; from the concentrated solution chloride of ammonium will precipitate vanadiate of ammonia, which is converted into vanadic acid, as above.

Vanadium is always mentioned, when stating the results of an analysis, as vanadic acid.

### XCII.—TANTALITE AND TANTALUM.

Tantalum and tantalic acid are substances of which we do not at present possess a sufficient knowledge to separate them entirely and with accuracy from the others with which they are combined. Even the atomic weight of tantalum has not yet been properly decided upon, and few quantitative analyses, in which this element is isolated, can be regarded as correct. Professor Rammelsberg regards 2607.43 as the equivalent of tantalum; while professor Koehler states it to be 1153.72. The difference is sufficiently marked to show how much all the most necessary characteristics of substance in quantitative analysis are hidden in mystery.

Tantalic acid much resembles the titanic acid, and like this is difficult to dissolve in acids. When united with alkalies it is soluble, and after acidulating the solution, zinc will precipitate white tantalic acid.*

As seen from the subjoined analysis, it is sufficient to state, in the result of the investigations, the quantity of metallic acids found.

Hermann found the tantalite from Middletown in Connecticut to contain:—

| | |
|---|---|
| Metallic acids | 78.22 |
| Oxide of tin | 0.40 |
| Tungstic acid | 0.26 |
| Protoxide of iron | 14.06 |
| Protoxide of manganese | 5.63 |
| Magnesia | 0.49 |
| | 99.06† |

* Koehler, *Chemie in techn.* Beziehung, p. 234.

† Rammelsberg. *Repertorium des Chem. Thls. der Min.* Number iv. p. 219.

### XCIII.—BLACK TELLURIUM ORE AND TELLURIUM.

Black tellurium consists of tellurium, lead, silver, and sometimes gold. A variety analyzed by Klaproth contained as much as 9 per cent. of the latter, though, as seen from the following analysis by professor Gustavus Rose, a specimen from the Altai Mountains contained none, the result being:—

| | |
|---|---|
| Tellurium . . . . . | 38.37 |
| Lead . . . . . . | 60.35 |
| Silver . . . . . . | 1.28 |
| | 100.00 |

Nitric acid, aided by heat, dissolves it, leaving gold, if that metal is present.

The amount of tellurium, after ascertaining its presence, is best determined by the loss, as even in its metallic state it is very volatile at a red heat.

Lead and silver are separately precipitated, as already described.

Tellurium is a very rare metal, and therefore one, the ores of which it will, in very few cases, be necessary to examine.

## XCIV.—MUD,* CONTAINING SULPHURIC ACID AND SELENIUM; A MIXTURE OF SULPHATE OF LEAD WITH SILICA, &c.; SELENIUM, SELENURET OF MERCURY, AND SELENIATES AND SELENITES. TREATMENT OF SELENIUM.

The dried mass should be intimately mixed with about one quarter of its weight of nitre, and then covered with half its weight of a mixture of equal parts of sulphuric acid and water, and with it heated until it has become perfectly white, and all the nitric acid has been evolved.

We then dilute with water, and filter. The filtrate contains, besides lead, iron, copper, and mercury, all the selenium in the shape of selenic and selenious acids. About half as much chloride of sodium as the mud originally contained should now be dissolved in it, when we boil it down to about one quarter of its volume. By the thus liberated hydrochloric acid all the selenic acid is converted into selenious acid.

On cooling, the liquid is poured off from the

* Produced by stamping and washing the ore. Such seleniferous lead ores are common, for instance, on the Hartz Mountains in North Germany.

deposit of sulphate of potash and chloride of sodium. The latter are then washed a few times, and the liquid portion saturated with sulphurous acid gas, prepared from a mixture of charcoal powder and concentrated sulphuric acid.

The precipitation of the selenium is facilitated by digesting for a long time, and towards the end of the process, boiling for about a quarter of an hour. This causes it to blacken and to collect as a dense and heavy powder.

The filtrate from this is again to be heated to the seething point with raw muriatic acid, and a second time to be treated with sulphurous acid, in case it should still contain selenium.

The selenium procured contains, besides trifling admixtures of lead, copper and iron, in particular mercury. On distilling, the former remain as selenurets.

To remove the mercury, the distilled selenium is dissolved in aqua regia, the chief excess of acid evaporated off, so that no nitric acid remains, and the liquid saturated with carbonate of soda, evaporated to dryness, and the saline residue heated to redness to evolve the mercury.

After dissolving again in water, we boil with

hydrochloric acid, and again precipitate the selenium with sulphurous acid.

We may also mix the calcined saline mass with about an equal proportion of sal-ammoniac, and placing it in a retort, apply heat, until most of the chloride of ammonium is sublimated, and the selenium is thus reduced.

Selenium is characterized by the peculiar odor of its fumes, when reduced, e. g. on charcoal by the blowpipe. It strongly resembles the smell of horseradishes.

### XCV.—HYDROCYANIC, OR PRUSSIC ACID.*

To ascertain the amount of hydrocyanic acid suspended in a solution, in an uncombined state, a weighed portion of the latter is mixed with nitrate of silver, while constantly keeping it in motion, until it ceases to become cloudy, and the well-known odor of prussic acid—which, amongst other things, is the same as that of peach-kernels and wild-cherry leaves—is no longer perceptible.

The precipitated cyanide of silver should be

---

* This acid might perhaps not be considered as one which should be mentioned in this book. It is, however, one of much interest, and is frequently combined with inorganic bases.

collected on a filter, previously dried at 250°, and weighed, and then itself dried at the same temperature and weighed.

To ascertain the quantity of hydrocyanic acid in the aqua amygdalarum amarum and aqua lauro cerasi, these must be mixed first with ammonia, then with a soluble silver salt, and afterwards with nitric acid.

In all cases, the hydrocyanic acid should be weighed as cyanide of silver, the formula of which is AgCy, while that of hydrocyanic acid is HCy; therefore, it is a matter of little trouble to ascertain the quantity of cyanogen in cyanide of silver, and from that to calculate the amount of prussic acid. Cyanide of silver contains 19.42 per cent. of cyanogen, which consists of two atoms of carbon and one of nitrogen, and hydrocyanic acid contains 89.7 per cent. of cyanogen. All this being known, it is very easy to calculate for each specific case, according to the directions given in the introduction, how much hydrocyanic acid can be formed, and was so originally, from the cyanogen contained in a given amount of cyanide of silver, which amount was ascertained by the analysis.

XCVI.—IODIDE, BROMIDE, AND CHLORIDE OF SODIUM,* AND IODIDE AND CHLORIDE OF POTASSIUM.

Nitrate of the protoxide of palladium is mixed with the solution of the compounds, and thus all the iodine precipitated as brownish black iodide of palladium. After twelve hours, the precipitate is collected on a weighed filter, dried over sulphuric acid, and weighed.

From the filtrate, bromine and chlorine are precipitated by nitrate of silver. The filtered and washed precipitate should be melted. A portion of it ought to be weighed off in a tube, with a bulb (see Fig. 8), and then melted in a slow current of dried chlorine gas, until the vapors of bromine are no longer perceptible, and no longer any alteration in weight takes place. Before weighing, and thus ascertaining the amount of bromine by the loss, it is necessary that all the chlorine gas should be removed.

A more simple plan is to pour some water and a

---

* These three combinations of sodium occur, for instance, in all sea salt. The very characteristic reaction of iodine with starch is too well known to require any description here.

few drops of hydrochloric acid over the molten and weighed mixture of bromide and chloride of silver, and to lay a piece of zinc upon it. After twenty-four hours, the silver will be entirely reduced. It should be crumbled up, so to speak, washed first with water acidulated with muriatic acid, then with pure water, and heated to redness and weighed.

The difference between the equivalent of chlorine and that of bromine, is to the equivalent of bromine as the discovered difference of weight between the employed quantity of chloride and bromide of silver, and the amount of chloride of silver, which the reduced silver would produce, is to the quantity of bromine sought.

For example, 200 parts of a mixture of 100 parts of chloride of silver, and 100 parts of bromide of silver, give, on reduction, 132.73 parts of silver, which would produce 176.31 parts of chloride of silver.

The difference between the equivalents of chlorine and bromine is 556.34. The difference of weights between the compounds, given in the last but one paragraph, is, 23.69. Therefore,

$$556.34 : 999 = 23.69 : X \ (=41.99 \text{ bromine}).$$

In the same indirect manner, the amount of iodine contained in a mixture of chloride of sodium and iodide of sodium, or bromide of sodium, may be ascertained.

In a mixture of chloride and iodide of potassium or sodium, the quantity of iodine may also be ascertained, by mixing their solution with one of sulphate of copper (in sulphurous acid), by which all the iodine is precipitated as white iodide of copper, which should be washed, dried at 250° F., and weighed.

This treatment is also applicable in a merely approximative separation of iodine and bromine.

## XCVII.—PURE CHLORIDE OF SODIUM.

To remove the inclosed water, the perfectly pure crystallized table salt should be intensely heated. After that it should be weighed, dissolved in water, and the chlorine precipitated by nitrate of silver. The chloride of silver, thus produced, ought then to be treated as directed in Art. XXX., and weighed.

To determine the quantity of sodium, a distinct weighed portion is placed in a platinum crucible, of known weight, and covered with concentrated sulphuric acid. Afterwards, slight heat is applied, until all the hydrochloric acid thus formed is evolved. The superfluous sulphuric acid should be driven off with great care, and the residue of sulphate of soda gradually heated to redness, while, to destroy the acid salt, a piece of carbonate of ammonia is held in the crucible. From the weight of the sulphate of soda, the amount of the sodium is determined.

Sulphate of soda consists of

| | |
|---|---|
| Soda . . . | 43.64 |
| Sulphuric acid . | 56.36 |
| | 100.00 |

### XCVIII.—SULPHATE OF SODA.

After determining the water of crystallization, a weighed quantity of the calcined salt should be dissolved in water, the solution diluted, and the sulphuric acid precipitated from it by chloride of barium.

When, on application of slight warmth, the precipitate has collected, it ought to be removed by filtration, washed with hot water, dried, and, as far as possible, removed from the filter. The latter is then to be entirely burned up, and, with the precipitate, heated to redness, and weighed.

The amount of sodium is deduced from the difference of weight between the sulphuric acid in the sulphate of barium and the weight of the whole.

### XCIX.—CHLORIDE OF SILVER AND PREPARATION OF PURE SILVER.

Chloride of silver is prepared by dissolving a weighed quantity of silver in nitric acid, and precipitating with dilute hydrochloric acid. It may then be treated as in Art. XXX., or otherwise, by heating a weighed portion of the previously molten chloride of silver in a bulb tube (Fig. 8, p. 108), over a spirit-lamp, in a current of hydrogen, until entire reduction is produced, and weighing the remaining pure silver. This plan is sometimes adopted to prepare the pure silver required for the quartation of gold ore.

Bodeman gives the following directions for the preparation of pure silver, for the same purpose:—

A silver coin is dissolved in nitric acid, of moderate strength and perfect purity, and the solution diluted with, at least, an equal proportion of water. After some time has been given for the gold to settle—most silver, particularly the Mexican, contains more or less—the solution is to be passed through a double filter, and then a clear solution of chloride of sodium mixed with it. The chloride of silver should be several times washed with clean water, until this can be decanted in an unaltered state, and without exhibiting any sign of the presence of copper. The chloride of silver is then to be reduced, by placing iron or zinc rods in it, and adding sulphuric acid in a sufficient quantity to cause a constant production of hydrogen, and in this state leaving it for a few days. When the process is concluded, the remainder of the zinc or iron is removed, and the reduced silver washed first with very dilute sulphuric acid, and then with pure water. After that, it should be melted with some calcined borax mixed with potash or nitre, and poured out into water to produce little granules,* as in this shape it is best adapted to perform its part as one of the reagents in an assaying office.

### C.—RAW TABLE SALT; ROCK, OR SEA SALT.

A weighed portion of the moist (hygrometric) salt is first dried for some time at 212° F., then covered and heated to about 570°, and thus the amount of water, both hygroscopic and of crystallization, determined.

---

* Bodeman's *Elements of Assaying* (German). Clausthal, 1845, p. 86.

To ascertain the quantity of sulphuric acid, the salt is dissolved in water (when some insoluble parts remain), the solution acidulated with hydrochloric acid, and precipitated with chloride of barium. Sulphate of baryta contains 0.34368 of sulphuric acid.

The lime is determined in a different larger portion of the solution, by precipitating it with oxalate of ammonia. When, aided by digesting heat, the oxalate of lime has collected, we filter, heat to redness, and weigh, as directed in Art. VI.

The filtrate from the lime should be concentrated by evaporation, and mixed first with ammonia and then with phosphate of soda. The ammoniaco-phosphate of magnesia should be left about twenty-four hours to collect and settle, when we filter, wash with ammonia, and proceed as in Art. VI.

The percentage of potash is generally very small. It is to be determined by concentrating a very large quantity of the solution of the salt, until a considerable portion of the chloride of sodium is deposited in crystals. From the mother-lye the potash is precipitated by chloride of platinum, as in Art. I.

Bromine is detected by introducing into a large

quantity of the solution some chlorine gas, adding ether, and shaking. The bromine is taken up by the ether, and colors it yellow. It should then be saturated with ammonia.

Iodine is discovered by mixing the mother-lye with a little starch, and, drop by drop, adding diluted chlorine water.

For determining of iodine and bromine quantitatively, see Art. XCV.

### CI.—DEPOSIT OR CRUST FORMED IN THE INTERIOR OF SALT VATS.*

This substance generally contains chloride of sodium, sulphates of soda, lime, and magnesia,

---

* Quarizius, in the work already quoted, mentions the following, with regard to this substance, which, in German, is called *Pfannenstein*.

"Whether the saline solution (either sea-water or produced by dissolving rock-salt) be previously freed from the foreign salts or not, a crust on the interior surface of the vat is produced by boiling. When this process is concluded, it is necessary to scrape it off, as it injures the vats, and is detrimental to any future evaporating process.

* * * * * *

"In some salt-works it is only sold as manure, and sometimes even thrown away. In a pecuniary point of view this is, however, not advisable, as, in general, this carbonates of lime and magnesia, oxide of iron, and sometimes, also, sulphate of potash and manganese.

A weighed portion is nearly heated to redness, and in this manner the quantity of water ascertained.

Another quantity should be pulverized, and dissolved in seething water, when the carbonates of magnesia and lime remain insoluble, and are removed by filtration. They must then be washed, and treated as in Art. VI. Sometimes they also contain iron and manganese, as peroxides.

The filtrate we mix with sal-ammoniac, and precipitate the lime with oxalate of ammonia. (See Art. VI.)

To the solution, filtered off from this precipi-

---

deposit, besides gypsum, silica, lime, oxide of iron, and dirt, contains common salt, Glauber salts, and chloride of magnesium, in considerable quantities, and may, therefore, be advantageously employed in concentrating a poor saline spring or solution, as well as in the preparation of Glauber salts, Epsom salts, and muriatic acid. For what purposes, however, this crust is best adapted, a chemical analysis of it can alone decide, as the original solutions, and hence, also, the mother-lyes, as well as these deposits, are of different composition in different cases."

tate, we add caustic ammonia, and precipitate the magnesia, as in Art. VI., with phosphate of soda.

A third portion of the compound we dissolve in hot water, acidulated with hydrochloric acid, and throw down the sulphuric acid with chloride of barium. (See Art. XVII.)

A small portion of the substance is to be dissolved in water, containing nitric acid, and the chlorine extracted by nitrate of silver. (See Art. XXX.)

The amount of sodium and soda is calculated from the loss.

To discover the presence and quantity of sulphate of potash, a large proportion of the substance should be ground up, and boiled with an excess of hydrate of baryta. The solution being filtered, the lime and baryta are precipitated by a mixture of caustic ammonia and carbonate of the same. The solution is then filtered off, evaporated to dryness, the residue heated to redness, and mixed with chloride of platinum.

In this manner, also, all the soda which had been combined with the sulphuric acid may be obtained as carbonate.

## CII.—COMMON SPRING OR WELL WATER, MINERAL WATERS, SEA WATER, AND THE MOTHER-LYE IN SALT-WORKS.

It would not be compatible with the size and purpose of this volume to enumerate any of the results of the analyses of the many and various mineral springs of even the United States, which so materially differ from one another in composition, and in the amount of their constituting ingredients. From good analyses of sea water—which does not vary much, except where local causes, such as the proximity of rivers or large deposits of some more or less soluble mineral substance may alter its character—it appears that the greatest proportion of salt in sea water is 3.77 per cent. and the smallest 3.48. Sparmann, by various experiments, has shown that those salts which give to sea water its bitter and nauseating taste, predominate in the portions of the ocean nearest to the surface, while the deeper we go, the more does the taste approach to that of chloride of sodium; in other words, become purely salt.

In testing a mineral spring on the spot, we cannot expect to obtain more than a very superficial idea of its composition, and even for that it is necessary to reduce it in bulk. The most common and easily distinguished ingredients are, carbonates of iron, lime, magnesia, and soda, chloride of sodium, potash, sulphuretted hydrogen, and free carbonic acid gas. Before proceeding to the quantitative analysis, some brief directions for a cursory investigation on the spot, may be of some service.—The presence of hydrosulphuric acid gas is betrayed, without any farther treatment, by its strong odor, resembling that of rotten eggs.—Carbonate of iron is already detected, if in large quantities, by the color of the chalybeate water, and also, even if in smaller quantities, by the rust-colored deposit produced by boiling down a portion of the water. The most delicate test for iron is cyanide of potash, which, even in human perspiration, can show the presence of that element. It produces a pink color, and, to make sure of its having effect, it is well to add a drop of acid. That this must not produce the reaction of iron itself, with that reagent, is a matter of course.—Lime is precipitated as an oxalate, magnesia as an ammoniaco-phosphate, and potash, from a greatly condensed quantity of the water, as in Art. I.—The chlorine produced by the decomposition of the chloride of sodium, is extracted and thrown down by nitrate of silver.—The presence of soda is observed by evaporating a considerable quantity of the water to dryness, and afterwards pouring alcohol over it, and setting it on fire, when, if soda exist in the water, the flame will receive a bright yellow color.—Free carbonic acid gas is detected before evaporating, by the white precipitate, formed with hydrate of lime.

For the quantitative analysis, it is presumed that the operator has at disposal an indefinitely large quantity of water, so that, in determining most of the ingredients, separate portions may be used. For those substances which occur in large quantities, little water may, of course, suffice, though for those which are not so abundant, a great deal of water must be employed.

1. The *specific gravity* of the water ought first to be ascertained, so that we may be able to calculate how much a given quantity may weigh, and thus know the weight of each different portion we may employ.

2. *Carbonic acid gas* and *hydrosulphuric acid gas* can only be determined on the spot by the following means: the former free acid, by placing a weighed or measured quantity of the water in a bottle. This should be capable of being closed, and

contain a clear mixture of chloride of lime, or, better, chloride of barium, with an excess of ammonia. After twenty-four hours, we filter (access of atmospheric air being avoided), wash with ammoniacal water, and, after drying, expose the precipitate to a moderate red heat. Should the water contain other substances, which ammonia will precipitate, this must not be overlooked.

CARBONATE OF BARYTA.

| | |
|---|---|
| Carbonic acid . . . | 22.36 |
| Baryta . . . . | 77.64 |
| | 100.00 |

CARBONATE OF LIME.

| | |
|---|---|
| Carbonic acid . . . | 71.43 |
| Lime . . . . . . | 28.57 |
| | 100.00 |

The sulphuretted hydrogen is best precipitated as sesquisulphuret of arsenic, by mixing the water which contains it in a closed bottle or vial, with a muriatic solution of pure arsenious acid in excess, giving time to collect, and filtering on a weighed filter. We then wash with cold water, dry at 212°, and weigh.

SESQUISULPHURET OF ARSENIC.

| | | |
|---|---|---|
| Arsenic | . . . . . | 60.95 |
| Sulphur | . . . . . | 39.05 |
| | | 100.00* |

3. The *sum total of the soluble matter* is determined by evaporating a known quantity of the water to dryness, and carefully heating the residue to 212°. If the solution contain much chloride of magnesium, the result will be more or less inaccurate, as it will, in part, be converted into hydrochloric acid and magnesia. To avoid this, a weighed quantity of calcined carbonate of soda is to be dissolved in the water.

4. The *carbonates of iron, manganese, lime, and magnesia,* dissolved in the free carbonic acid, are obtained as a deposit, by boiling a considerable quantity of the water in a retort for the space of an hour. After filtering, the precipitate is dissolved in hydrochloric acid, the peroxide of iron precipitated with ammonia, and protoxide of manganese, lime, and magnesia, separated as in Art. XVIII.

5. The *silica* remains insoluble when the solid

* See Frezenius's *Quantitative Chemical Analysis.*

residue, produced by evaporation, is dissolved in water acidulated by hydrochloric acid. If the mineral water contain carbonate of soda, hydrochloric acid should be added before evaporating. If it contain sulphate of lime, a corresponding large amount of water must be employed to redissolve it.

6. *Chlorine* is precipitated by nitrate of silver, after adding some nitric acid to the water. The precipitate is treated as in Art. XXX.

7. *Bromine* and *Iodine* can only be detected in a very large quantity of water or in mother-lye. Their amount is ascertained as in Art. XCV.

8. *Sulphuric Acid* is precipitated by chloride of barium, after slightly acidulating the water with hydrochloric acid. See Art. VII.

9. *Potash and Soda.*—We boil the aqueous solution down to about one-half its original volume, and, without filtering, mix it with an excess of water of baryta, and, on cooling, with carbonate of ammonia, by which all sulphuric acid, lime, and magnesia, as well as the excess of baryta, are precipitated. The filtrate from them is to be mixed with hydrochloric acid, and evaporated to dryness. The residue, a mixture of the chlorides of potassium and sodium, is carefully exposed to a nearly

red heat. Potash and soda are then separated, as in Art. I.

10. *Carbonate of Soda.*—The water should be boiled for some time, and the deposit of carbonated earths removed by filtration. The filtrate is divided into two equal parts. The one should be slightly acidulated with nitric acid, and the chlorine precipitated from it by nitrate of silver. The second portion is mixed with a small excess of hydrochloric acid, and evaporated to dryness. The residue is then heated very nearly to redness, dissolved in water, and the chlorine precipitated with solution of silver. The excess of weight of the last produced chloride of silver over the first, corresponds with the quantity of carbonate of soda which was present.

11. The *lime* is precipitated from the filtrate produced in 4, by oxalate of ammonia, after first adding some sal-ammoniac. (See Art. VI.)

12. *Magnesia.*—The water filtered off from the lime precipitate we concentrate by evaporation, and, on cooling, add concentrated ammonia to it, and precipitate the magnesia with phosphate of soda. (See Art. VI.)

13. *Lithion.*—We proceed, as in 9—the best is with mother-lye—add phosphate of soda to the fil-

trate, evaporate to dryness, and dissolve the residue in the least possible quantity of water, when phosphate of soda-lithion remains.

Or we add an excess of carbonate of soda to the mother-lye, evaporate to dryness, macerate the residue in hot water, add phosphate of soda to the filtered solution, and again evaporate to dryness.

14. *Strontia* should be sought for in the deposit, or ochre produced in calcareous waters.

15. *Phosphoric Acid and Phosphates.*—A large quantity of water is boiled down and reduced very much in bulk, and then mixed with ammonia. The precipitate is removed by filtration, dissolved in hydrochloric acid, and reprecipitated with ammonia. The phosphoric acid is to be determined in it, as in Art. XXIII.

16. *Arsénic Acid*, combined with lime or iron, may also be found in the calcareous deposits of some waters. The apparatus of Marsh is employed. (See Appendix.)

17. *Fluorine*, as fluoride of calcium, also occurs in some deposits, as well as sometimes in the last ammoniacal precipitate of 15. A portion of this, after being dried, should be placed in a crucible, and moistened with concentrated sulphuric acid, and, by means of a plate of glass covered with

wax—which is removed on some places by a pointed piece of wood, *i. e.* engraved—tested for fluorine.

### CIII.—SOILS.

The common ingredients, but which change very much in quantity, are salts of chlorine, sulphuric acid, phosphoric acid, silicic acid, carbonic acid, nitric acid, with potash, soda, ammonia, lime, magnesia, alumina, protoxide of manganese, protoxide of iron, beside quartz sand and organic matter, consisting of the remains of plants, and the products of their decomposition, the humic compounds.

A portion of these substances are soluble in water.

Others are insoluble in water, but soluble in diluted acids, as, for example, the carbonates and phosphates of lime and magnesia.

The remainder are insoluble, even in diluted acid. To these belong the quartz sand, and the particles of felspar, mica, and hornblende, originating from the disintegration of rocks.

The soil to be examined should be taken from different parts of the field or surface of the ground; then crumbled up, dried in the air, and thoroughly mixed up together.

It is most convenient to use different portions of the earth for the determination of most of the separate component parts.

1. *Water.*—A weighed quantity of the air-dried earth should be dried at 212°, until no longer any change in the weight takes place. In this manner the hygroscopic moisture is determined.

To ascertain the quantity of chemically combined water existing, for instance, in the salts and in the clay, the soil might be heated up to 400° or 550°, by which, however, the ammonia is also lost.

2. *Organic Matter.*—The dried soil should be heated to redness, then moistened with carbonate of ammonia, and again exposed to an almost red heat. The diminution of weight shows the whole amount of the burnt organic matter, including ammonia and nitric acid also.

The quantity of nitrogeniferous organic matter can only be accurately ascertained by an elementary analysis, the amount of ammonia and nitric acid in the soil being then also brought into consideration.

Certain organic substances, such as fats and resins, can only be extracted from the soil by means of hot alcohol or ether.

The humic compounds are capable of being separated from the rest by boiling the soil with a solution of caustic potash. From the filtered brown solution they are extracted, though not entirely, as a brown precipitate, by saturating with hydrochloric acid.

3. *Ammonia.*—The soil is distilled with a solution of caustic soda, and the ammonia collected and determined, as in Art. II.

4. *Nitric Acid.*—We must be satisfied with a merely qualitative detection. The earth should be macerated in water, and the solution produced, filtered, and concentrated down to a very small bulk. We then mix it with about one quarter of its volume of pure concentrated sulphuric acid, and add, drop by drop, a solution of sulphate of iron, which, if nitric acid be present, will produce a dark-brown coloring. Another plan is, to mix the very concentrated solution, in a tube, with copper filings, and to add sulphuric acid, when, nitric acid being present, yellowish-red vapors of nitrous acid are produced.

5. *The parts soluble in water.*—A large quantity of the air-dried soil, say 15,000 to 30,000 grains, should be heated with water for some time, almost to a seething temperature. We then filter, and wash the precipitate with hot water. The

whole liquid should then be evaporated down to about its original volume, and then either weighed or measured, when we proceed to the investigations relative to the following separate ingredients, with weighed or measured portions of the solution: —

*a.* The *total amount* of parts of the soil which are soluble in water, we ascertain by evaporating to dryness.

*b.* The *sulphuric acid* is precipitated by chloride of barium, after acidulating the solution with hydrochloric acid.

*c.* To determine the *chlorine*, we first acidulate with nitric acid, and then precipitate with solution of silver.

*d. Silica.*—We mix the solution with hydrochloric acid, evaporate to dryness, macerate the residue in muriatic (acidulated) water, and filter off the silica.

*e. Lime, magnesia, alumina, and protoxides of iron and manganese,* may be contained in the filtrate produced in "*d.*"—They are to be separated, as in Art. XXIII.

*f. Potash and Soda.*—The solution should be mixed with hydrochloric acid, evaporated to dryness, the residue dissolved in a little water, mixed with an excess of hydrate of baryta, and digested.

The precipitate is removed by filtration, and from the filtrate baryta and lime precipitated by carbonate of ammonia. The solution, filtered off from them, can only contain potash and soda, which should be separated and determined as chlorides.

*g. Phosphated Alkali.*—Its presence is only possible where no lime exists. We precipitate it as ammoniaco-phosphate of magnesia. (See Art. VI.)

6. *The parts insoluble in Water, but soluble in diluted Muriatic Acid.*—We weigh off about 8000 to 16,000 grs. of the washed and dried residue produced in "5," and thoroughly mix it up, *inter se.* It is then placed in a retort, and mixed up with water, until it is converted into a thin pap. We then apply moderate heat, and gradually add some hydrochloric acid, until effervescence is no longer produced. Warmth should be applied for some time yet, while constantly stirring, after which we filter, and carefully wash the insoluble residue. The solution should be concentrated, and either by weighing or measuring, portioned out into different parts. If the soil should contain much organic matter it is necessary to expose it to a moderate red heat, before permitting the acid to act upon it.

*a. Silica.*—We evaporate the solution to dryness while adding a little nitric acid.

*b. Sulphuric Acid.*—In a known (weighed or measured) portion of the solution filtered from the silica, the sulphuric acid is precipitated by chloride of barium.

*c. Alkalies.*—Another portion of this solution is treated with hydrate of baryta, as in "5, *f*."

*d. Phosphates of lime, magnesia, alumina, iron, and manganese,* are determined in the main portion of the filtrate from the silica, as in Art. XXIII.

*e.* The *carbonic acid* of the carbonates can be determined in a separate portion of the earth, macerated in water, as in the soda test.

*f.* A casual amount of *copper* and *arsenic* occurring in the soil is determined by separate experiments. (See Art. XXIII.)

7. *The parts insoluble in dilute Hydrochloric Acid.*—A small portion of the residue of "6," say 100 to 200 grs., is to be heated with several times its weight of concentrated sulphuric acid, until most of the acid is driven off. By this the clay, in particular, is decomposed. The almost dry residue is digested with diluted hydrochloric acid,

the solution filtered off and analyzed, as above, only that no silica need be expected here.

The residue, which was insoluble in dilute muriatic acid, should be boiled with a concentrated solution of carbonate of soda, for a good while. In this way the silica is dissolved. The filtered solution should be saturated with hydrochloric acid, evaporated to dryness, and the silica thus obtained by macerating in water and again filtering.

The portion which is insoluble in carbonate of soda, can be a mixture of sand, felspar, and other minerals not capable of being decomposed by sulphuric acid. They are distinguishable by the aid of a microscope. For a farther separation, they should be treated as in the analysis of felspar.

The remaining larger portion of the residue of "6," before treating with sulphuric acid, can be approximatively separated (mechanically, by washing) into its component parts. To do this we mix it up with water, by means of a feather, and pour off the minute suspended particles, in particular clay, until nothing remains, except sand and fragments of felspar, etc.

## CIV.—ASHES OF PLANTS.

The ashes of plants contain the salts of chlorine, phosphoric acid, sulphuric acid, carbonic acid, and silicic acid, with potash, soda, lime, and the oxides of manganese and iron.

The mode of analysis varies according as the ashes contain as much or more phosphoric acid, than is necessary to form salts with the iron, manganese, lime, and magnesia which are present. The ashes of seeds contain more phosphoric acid, while those of the different varieties of wood, and herbaceous plants possess no more than will suffice for the formation of those salts.

Manganese does not occur in all ashes, and its presence must therefore be established by a separate investigation before proceeding to the analysis proper.

The following analysis are taken from the United States Patent office report for 1849.

| | INDIAN CORN GRAIN. | WHEAT GRAIN. |
|---|---|---|
| Silica . . | 0.850 | 5.63 |
| Phosphoric acid | 49.210 | 43.98 |
| Lime . . | 0.075 | 1.80 |

| | INDIAN CORN GRAIN. | WHEAT GRAIN. |
|---|---|---|
| Magnesia . . | 17.600 | 11.69 |
| Potash . . | 23.175 | 34.51 |
| Soda . . | 3.605 | 1.87 |
| Sodium . . | 0.160 | — |
| Chlorine . . | 0.295 | — |
| Sulphuric acid . | 0.515 | 0.21 |
| Organic acids . | 5.700 | — |
| Peroxide of Iron | —— | 1.87 |
| | 99.175 | 99.98 |

According to professor W. Sheppard, of Charleston, South Carolina, the ashes of cotton wool is composed of,

| | |
|---|---|
| Carbonate of potash (with a trace of soda) | 44.29 |
| Phosphate of lime (with a trace of magnesia) | 25.34 |
| Carbonate of lime . . . . . . | 8.97 |
| Carbonate of magnesia . . . . | 6.75 |
| Silica . . . . . . . . . | 4.12 |
| Sulphate of potassa . . . . . | 2.90 |
| Alumina . . . . . . . | 1.40 |
| Chloride of potassium, sulphate of lime, phosphate of potassa, oxide of iron (a trace) and loss . . . . . | 6.23 |
| | 100.00 |

Professor Emmons, of Albany, New York, states the ashes of the heart wood of some trees to be as arranged from his analyses in the following table.

| Component parts of the ashes of the heart wood. | Sugar-Maple. (Acer saccharium.) | Hickory. (Carya alba.) | White oak. (Quercus alba.) | Red beech. (Fagus ferruginea.) | Bass wood. (Filia Americana.) | Butternut. (Juglans cinerea.) | Iron wood. (Ostrya Virginica.) |
|---|---|---|---|---|---|---|---|
| Potash . . . . | 4.21 | 12.21 | 9.68 | 4.04 | 4.05 | 1.00 | 14.55 |
| Soda . . . . . | | 0.05 | 5 03 | 25.53 | 10.41 | 14.82 | 0.09 |
| Chloride of Sodium | 0.08 | 0.06 | 0.86 | 0.24 | 0.52 | 0.13 | 0.10 |
| Sulphuric acid . . | 1.03 | 5.26 | 0.26 | 0.62 | 0.27 | 21.43 | 0.38 |
| Carbonic acid . . | 33.33 | 33.63 | 19.29 | 24.59 | 17.96 | 4.48 | 20 14 |
| Lime . . . . . | 43.14 | 43.52 | 43.21 | 31.82 | 45.24 | 42.02 | 27.46 |
| Magnesia . . . | 7.24 | 4.00 | 0.25 | 1.44 | 7.44 | 4.00 | 4.40 |
| Phosphate of Iron . | 1.34 | } | | 0.40 | 1.30 | 3.41 | } |
| " of Lime | 5.09 | } 6.34 | 13.30 | 22.04 | 8.96 | 0.59 | } 23.10 |
| " of magnesia | 0.22 | } | | 0.02 | 0.04 | 0.28 | } |
| Organic matter . | 1.93 | | 7.10 | 2.80 | 2.00 | 3.20 | ? |
| Silica . . . . . | 0.55 | 1.30 | 7.30 | 1.60 | 1.40 | 5.40 | 0.40 |
| | 98.16 | 105.47 | 100.00 | 115.14 | 98.79 | 100.76 | 90.53 |

There being, as already remarked, some difference between the means of analyzing seed and wood, these are here treated separately.

1. *Ashes of Seeds* (*Grain, &c.*).—About 1000 grains of the air-dried (or dried at 212° F.) seed should be burnt to charcoal in a platinum crucible at a moderate red heat. The coal is pulverized, repeatedly macerated in water, and the filtrate then evaporated to dryness.

The remaining charcoal should be dried, and being placed in a platinum crucible, so long covered with a small quantity of concentrated nitric acid, that the whole of the coal is destroyed. The remaining saline mass must be mixed with a little clean sugar, moistened with water and heated to redness, to destroy any nitrates that may have been formed.

The salts produced are mixed up together, weighed, and decomposed in the carbonic acid apparatus, where, at the same time, their percentage of carbonic acid is ascertained from the loss.

The solution is to be filtered off from the silica, the chlorine in the filtrate precipitated by nitrate of silver, an excess of the reagent removed by hydrochloric acid, the sulphuric acid thrown down by chloride of barium, and the superfluous baryta extracted by a careful addition of sulphuric acid.

The filtrate should be evaporated to dryness, hydrochloric acid poured over the saline residue, and with it digested to alter the character of the phosphoric acid. The major portion of the muriatic acid is then evaporated off, and the solution filtered from the residue of silica. This, together with the previously obtained silica, should be dried at from 300° to 370° F., and weighed. The admixed carbon, originating with the sugar, should be removed by heating to redness in an open vessel. Its weight, as ascertained by the loss, should be deducted from the whole weight of the ashes.

From the saline solution iron, lime, magnesia, and also, if present, protoxide of manganese, are precipitated by ammonia as phosphates.

If the precipitate contain no manganese, it should be heated to redness, weighed, then redissolved in hydrochloric acid, digested with it for some time, and again precipitated by ammonia. From the precipitate the phosphates of lime and magnesia are extracted by acetic acid, and the residue of phosphate of iron weighed as such. The acetic solution should be nearly neutralized, and from it the lime precipitated by oxalic acid, the magnesia by an excess of ammonia as ammoniaco-phosphate of magnesia, while the amount of phosphoric acid, which was combined with the lime is determined by the loss.

If, on the other hand, manganese be present, the compound precipitate is not to be heated to redness and weighed, but immediately treated with acetic acid, and the weight of the remaining phosphate of iron ascertained. From the hot acetic solution the phosphoric acid should be precipitated by acetate of lead, and an excess of the reagent removed by hydrosulphuric acid. An excess of ammonia will then throw down the manganese as sulphuret, after which the lime is precipitated by oxalic acid, and the magnesia by phosphate of soda. The phosphate of lead should be dissolved in dilute nitric acid, the lead precipitated by sulphuric acid and alcohol, and from the filtrate the phosphoric acid by sulphate of magnesia and ammonia, and weighed as pyrophosphate of magnesia.

The solution, from which the phosphated earths and oxides of metals were precipitated by ammonia, must be boiled to drive off the free ammonia, and while hot mixed with a diluted solution of acetate of lead, so long as a precipitate is yet produced by it. From this precipitate the phosphoric acid is extracted in the manner already described, and from the filtrate a slight excess of the salt of lead removed by hydrosulphuric acid. By evaporating and slightly heating to redness the alkalies are procured as chlorides. The potash should be weighed as potassio-chloride of platinum, and the soda determined by the loss.

Since the phosphate of lime is not entirely insoluble in a solution containing sal-ammoniac, it is possible that some lime may be mixed with the alkalies, and if so, it should, after removing the excess of oxide of lead by sulphuretted hydrogen, be precipitated by oxalate of ammonia.

2. *Ashes of Wood, Herbaceous Plants, &c.*—Of herbs which contain a great deal of ashes, 1000 grs. are employed for the investigation of their contents, but of species of wood which are commonly less rich in inorganic constituents, and also of the gramineæ, whose ashes chiefly consist of silica, about 2000 grs. are taken. They should be burnt to charcoal in a platinum crucible, the coal extracted by water, dried, and thoroughly burnt to ashes in an atmosphere of oxygen.

The analysis only differs from the former in so far as these ashes contain a larger proportion of alkaline earths than is required to saturate the phosphoric acid. Therefore, at the first precipitation with ammonia, all the phosphoric acid is extracted.

In the filtrate, the lime, manganese, and alkalies are contained. The lime is thrown down by oxalic acid, the magnesia by phosphate of ammonia, an excess of this reagent being extracted by acetate of lead; the alkalies are determined as above.

## CV.—GUANO.

As a general definition, we may consider guano as a mixture of ammoniacal salts and earthy phosphates. In the Peruvian and other South American guanos, the ammoniacal salts predominate, while the African varieties contain a larger proportion of phosphates and less ammonia. The following instructions, though perhaps not going as much into detail as might be desired by some chemists, will be found to suffice for any cursory examination, and to meet all the demands for practical purposes.*

The substances to be looked for are :—

1st. Water, ammonia, ulmic, uric, and humic acids, all of which may be classed as volatile, and organic matter, separable at a low red heat.

2d. Fixed alkaline salts, such as sulphate of soda, chloride of sodium, and alkaline phosphates, separable by the heat of boiling water from the previous ash.

3d. Earthy salts, consisting of the carbonates

---

* I am indebted for these directions to Isaiah Deck, Esq., by whom they were published in the English "*Pharmaceutical Times.*" O. M. L.

and phosphates of lime and magnesia, separable by hydrochloric acid from the residue of above.

4th. Sand or silica insoluble.

A.—Calcine 100 grains in a capsule, at a low red heat, until all black particles are burnt away and a white ash is left; weigh; the loss is No. 1.

Good guano should lose from sixty to seventy-four per cent. of this volatile organic matter.

B.—Digest residue of A in boiling water, which dissolves the alkaline salts ; filter, dry, and weigh; the loss is No. 2.

Good guano should lose from four to six per cent. of these alkaline salts.

The phosphoric acid can be separated from this solution by adding sulphate of magnesia and ammonia, which precipitates it as ammoniaco-phosphate of magnesia.

C.—Digest the residue of B in hot hydrochloric acid ; filter and wash well ; weigh ; the loss is carbonate and phosphate of lime and magnesia, precipitated by ammonia as a gelatinous precipitate, which, on being dried and submitted to heat, should amount to at least fifteen to twenty-five per cent. of the weight of the guano used.

D.—The residue, after the action of the hydrochloric acid, when dried and ignited, is sand or silica.

In good guano it ought never to exceed four per cent.

*Other Proofs of Goodness of Guano.*—Good guano contains from twenty to twenty-five per cent. of urate of ammonia, insoluble in water; from eighteen to twenty-four per cent. of undefined animal matter; and from fifteen to twenty of earthy phosphates; leaving from thirty-one to forty-seven per cent. of matter soluble in water, exclusive of moisture.

Decayed or bad guano yields sixty or seventy per cent. of its weight to water, from the uric acid and animal matter having wasted, and the large quantity of moisture in it, often amounting to from twenty-five to thirty-five per cent.

Good Peruvian guano does not lose more than seven to nine per cent. at a heat of 212°, and this includes a little ammonia.

*Further Proofs of Good Guano.*—Fifty to seventy per cent. should dissolve in a hot solution of caustic potash, with a strong smell of ammonia.

Hydrochloric acid, added in slight excess to the

filtered solution, should produce a copious brown crystalline precipitate of uric acid.

Specific gravity ought to be from 1.60 to 1.75.*

### CVI.—GUNPOWDER.

A weighed quantity of well-ground gunpowder should be dried over sulphuric acid, and in this manner the amount of hygroscopic water determined.

By digesting with water, the saltpetre is extracted. The remaining mixture of coal and sulphur should be dried at 212° F., on a filter previously dried and weighed at the same temperature. The sulphur and carbon being weighed, the saltpetre is ascertained from the loss, or otherwise, by evaporating the filtrate and weighing.

To determine the quantity of sulphur, one part of the thoroughly pulverized gunpowder should be mixed with one part of anhydrous carbonate of soda, one part of saltpetre, and four parts of pure, decrepitated table salt, and, being placed in a platinum crucible, heated until the compound is burnt, and has assumed a perfectly white color. We then dissolve in water, acidulate with muriatic acid, and precipitate the sulphuric acid with chloride of

* See analysis of Guano, in Appendix.

barium, as sulphate of baryta, which contains 13.747 per cent. of sulphur.

The amount of charcoal is discovered by the loss. It cannot be determined by macerating the sulphur in boiling potash lye or a solution of sulphuret of potassium, since it is not always entirely converted into carbon, and therefore would only partially be dissolved. We might succeed better by extracting it from the sulphur with sulphuretted carbon ($CO_2$).

## CVII.—MODE OF TESTING THE MANGANESE OF COMMERCE.

This substance, if good, consists of little more than peroxide (hyperoxide, deutoxide) of manganese, is crystalline, and gives a black powder. When after drying it is heated to redness, it discharges no water, or at most a trace. Commonly, however, it contains foreign admixtures, and amongst them the hydrate of the sesquioxide of manganese is most frequently met with. To determine its amount of peroxide, that is, in other words, to ascertain the quantity of oxygen it is capable of engendering, there are different methods.

1. A weighed portion of the very finely pulverized substance is placed in the apparatus employed for the quantitative determining of carbonic acid,

and brought in contact with sulphuric and oxalic acids, when sulphate of protoxide of manganese is produced, while all the oxygen, which can be conceived as combined with the protoxide of manganese, is evolved as carbonic acid.

One equivalent of pure peroxide is equal to 544, and produces two equivalents of carbonic acid equal to 550.

Hence, 0.99 gramme* of the peroxide forms 1.00 gramme of carbonic acid. In the test, it is best to use a treble quantity of the compound of manganese, 2.97 grammes, and for that 2.5 grammes of neutral oxalate of potash, and afterwards to divide the quantity of evolved carbonic acid by 3. It gives the percentage of the peroxide of manganese in the impure substance.

2. The well-ground and weighed manganese is mixed with water in a retort, capable of being closed, and then some bright and weighed strips of copper are inserted, and hydrochloric acid added. A narrow tube should be attached to the retort. We then allow the contents to digest until

---

* One pound troy (5760 grs.) is equal to 373.246 grammes. The advantage appertaining to the decimal system has induced me in this case, and one or two others, not to state the quantities in grains. O. M. L.

the compound of manganese is dissolved, taking care, however, that no free chlorine is evolved. After this, we heat up to a seething temperature for a quarter of an hour, then allow it to cool, pour off the liquid portion, wash the residue of copper, first with very dilute hydrochloric acid, and afterwards with pure water, dry and weigh.

Two equivalents of copper, equal to 791 parts, take up, to form the chloride, one equivalent of chlorine, equal to 443 parts. Hence, 791 is to 443 as the dissolved quantity of copper is to "$x$" (the amount of chlorine sought). 1.22 gramme of pure peroxide of manganese produce 1.00 gramme of chlorine, and therefore cause the dissolving of 1.78 of copper.

### CVIII.—TEST FOR CHLORIDE OF LIME OF COMMERCE.

The bleaching powder of commerce is a varying mixture of hypochlorite of lime, chloride of calcium with hydrate of lime. If treated with an acid, the whole of the chlorine is liberated and discharged. To ascertain its quality, that is, to determine the quantity of effective chlorine which it contains, different methods may be adopted.

1. We dissolve 14 grammes of pure arsenious acid in potash lye, and mix with it so much water

that, altogether, we have the volume of 2000 measuring parts on the graduated pipette. Then 100 of these parts of the solution will contain 0.7 gramme of arsenious acid, and the solution of chlorine, which is required to convert this into arsénic acid, contains 0.5 gramme of chlorine, since one equivalent, equal to 1240 of arsenious acid requires two equivalents, equal to 886 of chlorine, to become arsénic acid.

We weigh off 5 grammes of the bleaching powder, rub it up with water, pour it into a cylinder, and add so much water, that the whole is equal to 200 of the already mentioned measuring parts on the pipette.

We then place 100 such parts of solution of arsenic in a beaker glass, dilute with water, add an excess of hydrochloric acid, and color the liquid with from one to two drops of a solution of sulpho-indigotic acid (an indigo solution with sulphuric acid).

We next admix so much of the newly shaken chloride of lime solution (which was placed in the graduated pipette) to the colored solution of arsenic that the color just disappears. The solution of chloride of lime contained 0.5 of chlorine.

Had we used 90 measuring parts of the solution

of chloride of lime, then the five grammes of chloride of lime would have contained 1.111 gramme or 22.22 per cent. of chlorine.

2. A weighed portion of the bleaching powder is placed in a retort, covered with water, and mixed, first, with an excess of protochloride of iron, which must not contain any perchloride, and then immediately with hydrochloric acid, after which some bright weighed strips of copper are inserted. We then boil, until the perchloride formed is again converted into the protochloride, wash off the copper, dry and weigh. (See Art. CVI.)

### CIX.—METHODS FOR TESTING THE POTASH AND SODA OF COMMERCE.

The different kinds of potash and soda which occur in commerce, contain very varying admixtures of foreign salts. Their percentage of carbonated alkali, from which alone their value is derived, varies from 40 to 95.

Potash, in particular, often contains chloride of potassium, sulphate, silicate and phosphate of potash, and carbonate, phosphate, and silicate of lime.

Soda frequently contains chloride and sulphuret

of sodium, sulphate, silicate, and hyposulphite of soda, and also hydrate of soda.

In testing their amount of carbonated alkalies, different methods are employed.

1. *By carefully saturatiny a weighed portion with diluted sulphuric acid of known powers of saturation.*—To prepare the testing acid we mix, for instance, 70 grammes of English sulphuric acid with 600 grammes of water.

We next weigh off 5 grammes of pure, anhydrous carbonate of soda, dissolve in hot water, and color with a little tincture of indigo.

From a graduated pipette, we very carefully, particularly towards the end, add some drops of the testing acid, until the solution has just assumed a red color, and lines drawn with the liquid on litmus-paper commence, on drying, to leave a red mark.

We then observe how many measuring parts of the acid have been used. After this, so much water is added to the whole amount of the acid, that just 100 measuring parts of this diluted acid are necessary to saturate 5 grm. of pure carbonate of soda. This testing acid should be kept carefully bottled and closed. It shows, in potash or soda, the amount of carbonated or caustic alkalies directly as the percentage amount, if we weigh off from the substance to be tested a quantity equal to 5 grm. of carbonate of soda.

100 parts of testing acid saturate 5.000 grm. of carb. soda
100 parts of testing " " 2.935 " soda
100 parts of testing " " 6.487 " carb. potash
100 parts of testing " " 4.421 " potash

Therefore, if we take 6.487 grm. of a potash, the parts of acid used will show directly the amount of carbonate of potash; if we take 4.421 grm. we obtain the quantity of anhydrous potash.

A greater amount of silicate of potassa in the potash, and of silicate and hyposulphite of soda and sulphuret of sodium in the soda, would make the result inaccurate.

2. *By determining the quantity of carbonic acid evolved.*—The water we first determine by slightly heating.

a. *Potash.*—From the anhydrous, impure compound we weigh off 6.28 grammes in the apparatus for the quantitative determining of carbonic acid, and discharge the carbonic acid.

The quantity of the evolved centigrammes of carbonic acid, divided by two, gives the amount of carbonate of potassa in the potash, for 3.14 grm. of pure carbonate of potash produce 1.00 grm. carbonic acid. For the sake of greater accuracy we take the double amount, equal to 6.28.

b. *Soda.*—We take 4.82 grm. of the anhydrous compound, that is twice 2.41, which is the quantity of pure carbonate of soda, from which the acid evolves 1.00 grm. of carbonic acid.

If a soda contain caustic soda, known from the fact that, after adding an excess of chloride of barium, the solution is alkaline in its reactions, we proceed in the following manner.

4.82 grm. of the anhydrous compound are rubbed up with three parts of pure quartz sand and about one-third of pulverized carbonate of ammonia, moistened with water, and after some time slightly heated to drive off all the water and ammonia.

On cooling, the whole mass is placed in the evolving apparatus, and treated as above.

# APPENDIX.

## MODE OF DETECTING ARSENIC IN A CASE OF POISONING.

### MEDICO-JUDICIAL PROCESS.

When a case of poisoning by arsenic is suspected, the poison should be sought for in the contents of the stomach and the intestines, in the substance of these organs themselves, in the liver, the milt, the lungs, the vomit, the substance which produced vomiting, and in the urine and excrements: in all these it should be sought for and its presence established. According to the character of the case, it may be found in one or the other of these organs or in several at the same time. It may also be necessary to examine the remains of suspicious victuals and the vessels in which they were contained, as well as vessels or papers in which the arsenic employed may have been placed. In cases where the body has been buried for a long time, and, together with the coffin, has completely decayed, it not unfrequently may become necessary to examine the surrounding soil for an amount of arsenic which may have been contained in the body.

It is necessary that, previous to a chemical investigation, the contents of the stomach and intestines, as well as the vomit, should be submitted to a careful ocular inspection. The mass should be spread out in a clean porcelain vessel, and turned over with new glass rods or spatula, and examined with the aid of a magnifying glass. Particular attention should be paid to little white, hard particles, or granules, which may be undissolved arsenious acid. These should be carefully collected with a pair of forceps. They ought chiefly to be sought for in the parietes and folds of the mucous membrane of the stomach and intestines. By diluting the contents with water, or better still with alcohol, and thus extracting the lighter organic matter, we are frequently enabled, with considerable ease, to obtain a larger amount of the heavy arsenical powder.

The object in all chemical operations in a judicial investigation is to procure the arsenic in substance, and, in fact, in its elementary form as metallic arsenic. In this shape only it is possessed of such very characteristic properties, that it cannot be mistaken for any other substances, and may yet, in particles too small to be weighed, be recognized with certainty. Hence, only in this form should it be laid before a court of justice, and be recognized as a proof of its presence. All other combinations in which it might be procured ought to be considered as valueless and unsatisfactory evidence. This isolation of arsenic in its metallic state, even in the smallest possible particles, too small to be weighed, is very easy and simple. The main difficulty which can occur in the whole operation is when, as commonly is the case, it is necessary to extract such a minute portion of the poison from the large mass of organic matter of an entire corpse, no matter in what form we may procure it.

It is most convenient for the progress of the chemical investigation to suppose three different cases:—

1. The arsenic, still in shape of the original substance of white arsenic, exists in the contents of the stomach and intestines, or in the vomit.

2. It is intimately and invisibly, or in a dissolved state, mixed up with these contents and the chyme. It is, therefore, no longer to be distinguished in substance, nor mechanically to be separated.

3. Stomach and intestines are empty, or according to the investigation no arsenic could be found in them. It has therefore been absorbed, and has passed into the blood or the substance of the various organs.

1. *The arsenic is still present in substance*, and can be picked out from the contents of the stomach, &c., or else extracted by washing. This case is the simplest. It is then only necessary to show that the substance found is really arsenic. This can be established by discovering the following properties. The granules or pieces should first be perfectly cleansed from all organic matter.

These particles are generally milk white, less commonly clear and semi transparent, hard and brittle.

A granule, however small, if placed in a narrow tube, closed at one end, and introduced into the flame of a spirit-lamp, will evaporate and be condensed again a little higher up along the sides of the tube, giving it a coating resembling flour. This, if examined with a microscope, particularly if held so that the rays of the sun can shine upon it, will be discovered to consist of glittering octohedral crystals.

A granule placed upon a redhot piece of charcoal will evaporate, producing vapors of a strong garlic odor. If placed on a redhot glass or porcelain it will evaporate without the garlic odor, because it is then not reduced to metallic arsenic. This odor is characteristic of the metal.

We may farther test it by inserting a particle in an open narrow tube, together with some small pieces of newly prepared charcoal, so that the tube is filled up about half an inch. The part of the tube containing the arsenic and coal, which ought to be the centre, should then be held in a flame in a horizontal position, but so that the part where the grain of arsenic is lying, is immediately outside of the flame. As soon as the charcoal is redhot this portion also is held in the flame and slightly inclined downwards, so that the vapors of arsenic passing through the redhot charcoal become reduced, when they settle above the coal, forming a sparkling but deep steel-colored ring, which reflects like a mirror, and is frequently termed the arsenic mirror. By heating this ring slightly, it can be driven higher up in the tube and caused to form again above the places originally occupied. Should it be thus driven to a wider portion of the tube, where the atmospheric oxygen can affect it, and here forced from point to point, a portion at least of the metal will become reoxidized and changed into the minute glittering and colorless crystals of arsenious acid. If the tube is cut off just below the metallic mirror and then heated, the characteristic smell of garlic may be observed.

This reduction of arsenic to its metallic state can be with equal convenience and greater certainty effected in the following manner: A granule is dissolved in water (acidulated with hydrochloric acid), by applying slight warmth, and then testing this solution in Marsh's apparatus, in the manner as is accurately described below.

Another proof of the white granules really being arsenic, is to heat one of them, together with a small quantity (about the size of a pin's head) of dry acetate of potash, in a small glass tube, closed at one end, when the peculiar and excessively offensive odor of *kakodyle* must be produced.

One or a few of the little grains are ground up in an agate mortar with distilled water, when the powder, together with the water (twenty to thirty drops) is poured into a little vial, and in it exposed to an almost seething temperature, till the powder is dissolved. A part of this solution is placed in a test tube, and mixed with a few drops of a solution of nitrate of silver, after which, one by one, some drops of very dilute caustic ammonia are added. This, if the substance was arsenious acid, will produce a heavy, bright-yellow precipitate of arseniate of silver. Another portion of the solution, mixed with a few drops of a clear solution of cuprum ammoniacale, produces a fine yellowish-green precipitate of arseniate of copper. A third part of the solution, mixed first with a few drops of hydrochloric acid, and then with several times its volume of an aqueous solution of hydrosulphuric acid, gives a vivid yellow precipitate of sesquisulphuret of arsenic, which is again dissolved by adding caustic ammonia.

Among all these reactions, the reduction of the arsenic to its metallic state, in either of the two methods given, is the characteristic and most necessary one, and, indeed, the only one which can be regarded as a clear proof. The others are simply confirmations of that test, and are only made when we have sufficient material at our disposal.

2. *The arsenic is no longer perceivable in substance*, no longer capable of mechanical separation, but is, either in a dissolved state, or at least intimately and invisibly, mixed with the contents of the intestines. In this more difficult, though fre-

quently occurring case, it is first necessary to dissolve by decomposing reagents, and totally to destroy the whole mass of organic matter of the contents of the intestines, of the vomit, the victuals, and even of the stomach and intestines themselves. Only when this has been effected is it possible to separate the arsenic with certainty.

Before commencing this operation, it is absolutely necessary carefully to test the reagents to be used, for a possible and not uncommon admixture of arsenic, even if they were purchased or prepared as chemically pure. Distilled sulphuric acid, hydrochloric acid, and zinc in particular, require such an examination. This is best performed by means of the Marshian apparatus, to be described below, and which is so invaluable, from the very fact that by it also the purity of the reagents themselves can be so accurately and easily determined. Without such a previous examination of these, which the chemist should mention in his protocol, the discovery of arsenic in a procedure of the kind must be regarded as utterly valueless, since the arsenic discovered may have originated with the reagents. It should also be mentioned in the protocol, that the investigations were conducted by means of new unused utensils and vessels. To make the whole perfectly sure, and to avoid leaving any doubt in the mind and conscience of the operator, it is advisable not to perform the examination in the common workroom of the laboratory. At any rate, a careful and thorough cleansing of it should precede the accurate investigations.

If, on such treatment, and after attending to all the necessary measures of precaution, arsenic is found, it should be remembered, that it may have been brought into the body by mere chance; in particular, by the previous use of certain medicines; as, for instance, the antimonials, the preparations of phosphorus, the acidum phosphoricum, sulphuricum, and muriaticum, which, from a careless preparation, may contain arsenic. Even the antidote employed in a supposed case of poisoning, the hydrated oxide of iron, may, in consequence of being badly prepared, contain some arsenic. The arsenic may also have been intentionally employed as a medicine, as in Fowler's drops, for example. When the corpse is dug up, it should be remembered that the soil which was in contact with the coffin must be tested for an amount of arsenic, as sometimes the earth, in particular if it be a ferruginous one, contains arsenic in detectible quantities, and may have communicated this to the body.

To alter and destroy the organic matter, preparatory to extracting the arsenic, according to its nature, different means can be adopted.

Into pap-like masses, such as the chyme and other contents of the stomach, and the excrements, chlorine gas is conducted until saturation takes place. This gas should be produced by the means of tested distilled sulphuric acid, and equally pure peroxide of manganese, and for greater purification should be passed through a high and narrow column of water. To aid the action of the chlorine it will be well to warm the organic matter. Afterwards, when it is entirely saturated by the gas, and has become coagulated and bleached, it should be heated almost to a seething point, to discharge the superfluous chlorine, and then the liquid, which contains the arsenic, filtered off through paper which contains no smalt, *e. g.*, Swedish filtering paper.

The stomach and intestines, with the contents, are cut up very small in a porcelain evaporating dish, and the whole mass then equally stirred up together. About one-third of it should be set aside in a clean, closed glass, to be used in case that, by some mistake, the remainder might be lost during the investigation. A moderately concentrated potash lye is then poured over the mass, and warmed with it, until the latter is entirely, or almost entirely, dissolved. But little potash is necessary for this, and therefore, to avoid an excess, it should only be gradually added. The solution ought then to be slightly acidulated with dilute sulphuric acid, by which it is caused to coagulate. As in the last paragraph, it should now be saturated with chlorine gas.

Another plan is to cover the minced organic mass with so much pure concentrated hydrochloric acid that its weight would be about equal to the weight of the whole organic matter, if the latter were dried. Enough water is also added to convert the whole into a thin pulp or pap. The evaporating dish containing it is then heated in a water-bath, and while stirring every five minutes, about thirty grains (not granules, but the weight of that name) of chlorate of potash, containing no lead, are strewn into the hot liquid, until this has assumed a light color, and has become a thin, homogeneous mass. After applying heat for some time longer, the liquid is left to cool, and then filtered through a wet filter of paper containing no smalt. The residue on the filter should be washed with hot water, until the washings are barely any longer of an acid nature. The whole of the fluid part ought then to be poured together in a porcelain evaporating dish, and in a water-bath evaporated to about one pound residue.

The liquid procured by either of these three methods, should be poured into a cylindrical glass or a retort, and a slow current of washed hydrosulphuric acid gas passed through it, until it has become perfectly saturated. By this all the arsenic is precipitated as sesquisulphuret of arsenic. The precipitation can be materially aided by heating the liquid, while introducing the gas, for about half an hour, to 120° or 140° Fahr., and then letting it cool while continuing the current of sulphuretted hydrogen. The fluid thus saturated by the gas, is set aside in a covered vessel for twenty-four hours. The deposit formed, even with a maximum amount of arsenic, generally possesses a dirty, indefinite gray or brown color.* Most of the liquid is poured off from it, and after that it is placed

* In a case of poisoning by compounds of lead, copper, mercury, or antimony, it would also contain these metals, and, therefore, possess a darker color, and require a separate treatment for them.

on a filter of fine Swedish paper, of the smallest possible size, and then washed out. The filtrate, before throwing it away, should, for precaution's sake, be saturated afresh with the gas, and set aside for some time in a closed vessel.

The precipitate, besides sesquisulphuret of arsenic, also contains some sulphuretted organic matter, which it is necessary entirely to destroy. This is best done by placing the filter with the precipitate in a large crucible of genuine porcelain or China ware, and to cover and digest it with concentrated nitric acid until all is converted into one homogeneous mass. The free nitric acid, of which, according as it may seem necessary, a larger quantity is added, should be neutralized by gradually admixing pure carbonate of soda. The whole is then evaporated to dryness. It is of importance that it should contain the required excess of nitrate of soda, a matter easily effected. We now expose it to a greater heat over a large spirit-lamp, though very gradually, until the salt melts. At first, it turns brown and black, and then, without fulminating, it becomes colorless, and melts to a clear liquid. By this proceeding, all the organic matter is burnt, and all the arsenic converted into arseniate of soda.

On the cold and stagnated saline mass in the crucible, we pour by degrees pure sulphuric acid in drops, and afterwards apply a mild heat, until, on adding more acid, the nitric and nitrous acid has been entirely driven off, and the mass been converted into acid sulphate of soda. Had, for the original oxidation of sulphuretted hydrogen, precipitate nitric acid, which contained hydrochloric acid, been used, then volatile chloride of arsenic would now be formed, and produce a loss of arsenic. It is, therefore, necessary to assure ourselves of the purity of the nitric acid and carbonate of soda.

The acid saline mass should be dissolved, while yet in the crucible, by the smallest necessary quantity of hot water. This solution is then placed in the Marshian apparatus (Fig. 11).

Fig. 11.

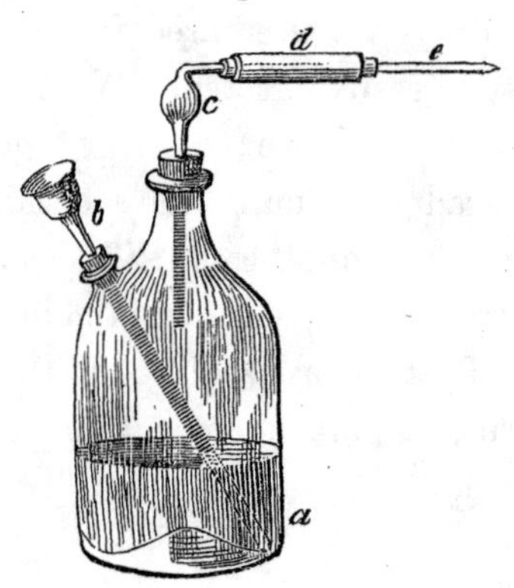

This apparatus has the following very simple construction: *a* is a bottle with two necks. It should be large enough to contain from a half to a whole pound of water. Into each mouth a new cork, with a hole passing through it, is very tightly inserted. Through the cork in the smaller aperture the funnel-topped tube *b* is passed, for the introduction of the solution, &c. Through the larger cork another tube is passed, having a bulb of an inch diameter at *c*. This serves to collect the spattering particles of the liquid, which from here, by means of the slantingly-cut end of the tube, again drop into the bottle. The opposite end of the tube is connected by a tight-fitting cork with the wider tube *d*, which is six inches long, and, to retain the moisture, filled with little pieces of pure chloride of calcium, which should have been previously melted, and must not contain any absolute powder. Into the other end of the tube *d*, a narrow tube *e* (at the utmost one-twelfth of an inch in diameter), made of glass containing no lead, is inserted. This should be of rather stout glass, and drawn out at the extreme end. It should be especially remarked that the introduction-tube, *b*, is indispensable. If the operator should have no two-necked bottle, he must insert a cork, with two perforations, into a single-necked one.

Some ounces of broken zinc are placed in the bottle, and the latter half filled with distilled water. The apparatus is then fitted together, and some concentrated distilled sulphuric acid introduced in small quantities by means of the funnel-tube *b*. This should be done very gradually, so that the whole does not become too severely heated, otherwise hydrosulphuric acid gas might be produced. The disengagement of hydrogen should be continued until all the atmospheric air is believed to be discharged from the apparatus, and hydrogen entirely to have taken its place. The narrow tube at *e* is then heated to redness with a Berzelius lamp, for at least half an hour, while at the same time, by occasionally pouring on more acid, the discharge of the gas is continued. By this, the zinc and acid are tested for a possible amount of arsenic. If these are pure, no change is produced at the point *e*. If they contain arsenic, a metallic mirror is produced on this spot, and both the acid and zinc are useless. The apparatus will then also require a thorough cleansing, and it is even better to exchange it for another new one. In a similar apparatus, the hydrochloric acid and the chlorate of potash are tested in the same manner. It is, however, necessary first to convert the latter entirely into chloride of potassium by melting. The saltpetre and hydrate of potash (for the third case) should be submitted to the same process, after converting them into sulphates by the action of sulphuric acid. Of all these substances too small quantities ought not to be employed. At least one ounce of each should be taken.

When the reagents have been thus tested and the operator has come to the conviction that they are absolutely free from arsenic, he may proceed to the investigation proper.

The apparatus should be entirely charged with hydrogen, and this gas kept continually producing while the narrow tube at *e* is maintained at a constant red heat. These preliminaries being arranged, the solution to be tested—and which must contain the arsenic originally existing in the corpse—is poured into the bottle by means of the funnel-tube *b*. Afterwards, about as much water is poured on, to rinse out this tube and remove the adhering particles of the solution. Care should be taken that no air is carried in by the liquids. If the solution contain arsenic, a deposit, at first of a brown color, but by degrees becoming possessed of reflecting properties, will be discovered, immediately beyond the redhot point. This increases in density, and with a large amount of arsenic will soon become an absolute intransparent metallic mirror. In the mean time, the current of hydrogen, emitted from the extreme end of the narrow tube, should be set on fire (the flame should not be too weak), and a piece of white and genuine porcelain held against it, on which then also a black or brown metallic spot of arsenic is produced. Often many of these may be caused to form. The reason why this takes place is, that if the portion of the tube which is heated to redness does not take up much room, then more or less of the arseniuretted hydrogen will escape decomposition, and produce these spots in the manner stated. Delfware, earthenware, or any kind of porcelain which is not genuine, cannot be used, since their glazing contains lead, and may, therefore, by producing a reduction of lead, cause the formation of dark spots, without the presence of arsenic.

If large quantities of arsenic are operated upon, and therefore many of these stains or spots are formed, these alone will be a sufficient evidence, as the reactions, specified above, can be produced with them. For this purpose, they should be dissolved by moistening them with a drop of nitric acid, and, by applying very mild heat, the greater part of the acid driven off. If, on the other hand, traces only of arsenic are present, the stain is so faint, that its character may frequently remain uncertain. A true proof only is the production of a metallic mirror or ring in the redhot tube itself. This ring must be capable of being evaporated by slightly heating the part of the tube where it has formed, and of being deposited anew in a cooler place. At the same time, the gas emitted from the end of the tube must possess the peculiar garlic odor.

When the arsenical ring has ceased to increase, and the flame no longer produces stains on porcelain, the operation should be stopped. It will be found much to the purpose, while the tube *e* is yet redhot, and in a softened state, to draw it out and melt it together (close it) at the redhot point. The other end may also be hermetically sealed in the same manner, and the metallic mirror, thus inclosed, placed in the hands of the judicial authorities.

If there be reason to suspect the presence of a

large proportion of arsenic, it is advisable not at once to take the whole quantity of the solution, but to divide it off into different parts, and also to let the tube *e* be a much longer one, so that the arsenical mirror can be produced at different points. By means of a file, the tube is then cut up into as many pieces as it contains rings. The portion with the most characteristic one is to be sealed at both ends, and laid aside, to be handed over with the written returns. To the rest, the tests given on page 265 are applied, and amongst them the great volatility, and the garlic odor, are most characteristic and decisive.

If, after applying the Marshian test, and even on heating to redness for an hour, no deposit is formed in the tube, and no trace of a stain is produced on porcelain, the absence of arsenic is proved, provided that the preparatory treatment was properly conducted, and the arsenic was not permitted to escape, from carelessness and inadvertence.

With this Marshian process, it is a circumstance of great importance that antimony, in the shape of oxide of antimony or antimonious acid, or altogether in its saline or dissolved state, under the same conditions as arsenic, will form antimoniuretted hydrogen, which, on being heated, produces a very similar metallic ring, and, on being burnt, causes stains on porcelain very much like the arsenic. This circumstance is peculiarly deserving of attention, since very commonly preparations of antimony, particularly tartarus stibiatus, are employed as internal medicaments, and would, therefore, in such cases, produce a metallic mirror, which at first sight might be easily mistaken for arsenic.

The methods to distinguish whether the ring be produced by arsenic or by antimony are very simple and decisive. The arsenic is characterized by the properties described, while antimony is possessed of the following marked qualities: The mirror of antimony is lighter in color and more iridescent, while the antimonial stains are blacker, and often possessing a shade of blue. Antimony is by far less volatile than arsenic, and, although the antimonial ring may also be driven from point to point, yet it requires a much more intense heat. Here one of the most characteristic differences is observable. It is, that the antimonial mirror, before it evaporates, will melt to small, glittering globules, at least distinguishable with the aid of a microscope; an intermediate state by no means observable with arsenic. The absolutely decisive distinction between the two is based on the fact, that the arsenic evaporates emitting the smell of garlic, while the antimony under the same circumstances possesses no odor whatsoever. By heating the portion of the tube containing the ring, while hydrogen is passing through it, a strong smell of garlic will be produced by the discharged vapors, if the mirror be arsenic, though they will remain odorless, if it is antimony. Besides these the following distinguishing reactions may be applied.

The stains on porcelain, caused by arsenic, disappear on touching (tipping) them with a concentrated and alkaline solution of hypochlorite of soda; those of antimony remain unaltered. Stains or mirrors of arsenic, if moistened with a drop of nitric acid, and heated, disappear, that is, are entirely and clearly dissolved. If a drop of nitric solution of silver be added, and a glass rod, moistened with caustic ammonia, be held immediately over the spot, without coming in contact, the drop will receive a yellow coloring, since a precipitate of arseniate of silver is produced in it.

It is true that the stains and mirrors of antimony also disappear on application of nitric acid, but the antimony is not dissolved by it, but only converted into white oxide, and this exhibits no reaction with the solution of silver. In a mixture of one drop of nitric and one of hydrochloric acid, the antimony is dissolved. If, then, the excess of white acid be carefully driven off, and some sulphuretted hydrogen water be dropped upon the spot, a fire red precipitate of sesquisulphuret of antimony is produced. If it had been arsenic, a lemon-colored precipitate would have been deposited.

If a current of hydrosulphuric acid gas be conducted through the tube, containing the metallic ring, and heat be applied, the metal will be converted into a sulphuret. If it is antimony, the sulphuret will be black, or partially orange-colored; if arsenic, the yellow sulphuret is produced. Both combinations are respectively characterized, not only by their differing in color, but also by their difference in volatility, the sulphuret of arsenic being much the more volatile of the two.

Besides this, the presence of antimony might have been detected on oxidizing the sulphuretted hydrogen precipitate, already described. For had the molten mass, before being treated with sulphuric acid, been dissolved in water, the antimony, in the shape of antimoniate of soda, would have remained insoluble.

3. *In the stomach and intestines, no arsenic is discovered;* and it is to be supposed, either that

it was discharged in the vomit or excrements, or that it was absorbed by the blood and bloodvessels of the body. In this case, the procedure is precisely the same as in the second, by subjecting the liver, milt, lungs, heart, or kidney to the investigations. If the bladder is found to contain urine, we should commence by examining this. It should, however, not immediately be placed in the Marshian apparatus, as the strong effervescence which it would produce in the liquid, would prevent our conducting the experiment in this manner. It is therefore necessary, first, slightly to acidulate it with hydrochloric acid, and then to introduce hydrosulphuric acid gas, after which we proceed as above in the second case.

Investigations of this kind become excessively troublesome, disgusting, and even painful, if we have to operate on corpses which have been buried for months or years, and have therefore passed over into the most offensive state of decomposition. In such cases it is often impossible to distinguish and separate the different organs, and we are obliged to submit to the examination the whole mass of the vital parts—which, from decay, have passed into one another, or, under peculiar local circumstances, have dried up together—and even the bones. It would be wrong, in such an instance, to immerse the corpse into a bath of chlorine water or a solution of chloride of lime, to deprive it of its stench, as then arsenic would be extracted and lost. Should the operator wish to disinfect it by chlorine gas, he must remember that, in preparing it, distilled sulphuric acid, containing no arsenic, should be employed. The vital parts, and in particular those which belonged to the abdomen, are carefully removed from the bones, and then treated as in the second case.

The following is another mode of treatment, which is peculiarly adapted to decomposing corpses which have only been buried for some months.

The tender (vital) parts are placed in a large evaporating dish of genuine porcelain, standing on a sand-bath, and there covered with moderately strong, tested nitric acid. By aid of heat and by stirring, they are then so completely destroyed a nd dissolved, thata homogeneous, pap-like mass is produced. This is then saturated with a concentrated solution of pure hydrate of potash, or carbonate of potash, and afterwards about as much tested and pulverized nitre admixed as the weight of the organic matter amounted to. The whole is then, while constantly stirring, evaporated to dryness, and afterwards in small portions place in a new, clean Hessian crucible, previously brought to a red heat. In this manner, all the organic matter is burnt, and the arsenic, if present, converted into arseniate of potash. In this case it is necessary, though not quite easy, to decide upon the right quantity of saltpetre. If too little be used, some organic matter will be left unburnt, and arsenic may be volatilized by the carbonated mass; while too much would be productive of inconvenience in the farther treatment. It is best to make some preparatory tests, by placing small quantities in little redhot crucibles, and observing whether, after the cessation of the detonations, the mass is left purely white. So long as it remains dark (carbonated) more nitre should be added.

This substance, which principally consists of carbonate, nitrate, and hyponitrate of potash, and may contain arseniate of potash, should be dissolved in the smallest necessary quantity of seething water. This solution, without removing the suspended phosphate of lime and silica by filtration, should be placed in a porcelain dish, and there be saturated in such a manner with concentrated distilled sulphuric acid, that of the latter a slight excess is produced. The saline, pap-like mass is then carefully heated, until the whole of the nitric and nitrous acids are discharged. This circumstance requires great attention. On cooling, the whole is mixed with a little water, and the liquid then poured off from the large quantity of sulphate of potash formed. The latter should now be washed several times with cold water, and the washings then mixed with the first liquid, when the whole of this solution is to be treated, as above, with hydrosulphuric acid gas. The precipitate then only requires to be oxidized with nitric acid, care being taken afterwards to remove all the acid by evaporation, before the liquid is submitted to the Marshian process.

In rare cases only may it be required by the judicial authorities to be made acquainted with the exact weight of the arsenic contained in a dead body.

In determining the quantity, the result can only be an approximation, since it is impossible to obtain the whole amount of arsenic from all the separate parts of a corpse, and to ascertain its weight. In such a case, where it is expressly desired, it is well to insert a closely-wound spire of thin copper foil, about two inches long, into the rather long tube *e*, and to weigh them together. The tube is then heated in two places, one near to the bottle for the formation of the mirror, the other at the somewhat distant part occupied by the copper. This latter will unite with all the rest of the arsenic, forming an alloy of the two metals. The increase in weight of the tube will then show the quantity of arsenic, which must, however, be calculated as arsenious acid.

A hundred parts of arsenic produce 132.0354 parts of arsenious acid.

ANALYSES OF COPPER OF COMMERCE.*

| | From Sweden. | From Mansfeld, in Prussia, by VON KOBELL. |
|---|---|---|
| Copper | 98.655 | 98.251 |
| Iron | 0.055 | 0.131 |
| Nickel | — | 0.236 |
| Lead | 0.751 | 1.092 |
| Silver | 0.226 | 0.135 |
| Potassium | 0.116 | — |
| Calcium | 0.095 | 0.107 |
| Magnesium | 0.033 | |
| Silicon | 0.048 | — |
| Aluminium | 0.021 | 0.048 |
| | 100.000 | 100.000 |

* *U. S. Report on Lake Superior*, by Foster and Whitney, p. 177.

THE AVERAGE COMPOSITION OF UNDAMAGED GUANO FROM PERU, IS, ACCORDING TO WAY, OF LONDON:*

| | | |
|---|---|---|
| Ammonia . . . | 17.41 | per cent. |
| Phosphate of lime . | 24.12 | " " |
| Potash . . . | 3.50 | " " |

besides water, organic matter, &c.

ANALYSES OF TIN BY KERSTEN.†

| | Altenberg in Saxony. | Peru. |
|---|---|---|
| Tin . . . | 97.83 | 93.50 |
| Lead . . . | — | 2.76 |
| Iron . . . | 0. 11 | 0.07 |
| Insoluble in hydrochloric acid . | 1.90 | 3.76 |
| | 99.84 | 100.09 |

* J. Higgins's *Agricultural Report of Maryland.*

† *Smithsonian Report on Recent Improvements in the Chemical Arts.* By Booth and Morfit, 1851.

### ANALYSIS OF SOFT PIG IRON, FROM BERGEN, BY FUCHS.*

| | |
|---|---|
| Iron . . . . . | 94.33 |
| Carbon . . . . . | 3.43 |
| Silicon . . . . . | 1.75 |
| Phosphorus . . . . | 0.37 |
| Sulphur . . . . . | 0.12 |
| | 100.00 |

### ANALYSES OF SELECT, PURE, AND PERFECT CRYSTALS OF ARSENICAL PYRITES FROM FREIBERG IN SAXONY. BY DR. C. J. B. KARSTEN.†

| | I. | II. | III. | IV. |
|---|---|---|---|---|
| Arsenic | 42.88 | 43.418 | 48.1 | 43.73 |
| Iron . | 36.04 | 34.938 | 36.5 | 35.62 |
| Sulphur | 21.08 | 20.132 | 15.4 | 20.65 |

* Karsten's *Eisenhuetten Kunde* (German), vol. i. p. 623.
† Ibid. vol. ii. p. 19.

ANALYSIS OF THE DRIED ALLUVIAL DEPOSIT OF THE NILE. BY LASSAIGNE.*

| | |
|---|---|
| Silica . . . . . . | 42.50 |
| Alumina . . . . | 24.25 |
| Oxide of iron . . . | 13.65 |
| Carbonate of lime . . . | 3.85 |
| Carbonate of magnesia . . | 1.50 |
| Ulmic acid and organic nitrogeniferous substances . . | 2.80 |
| Water . . . . . | 10.70 |
| | 100.00 |

ANALYSIS OF THE CRETACEOUS (ROTTEN OR PRAIRIE) LIMESTONE OF ALABAMA.†

| | |
|---|---|
| Carbonate of lime . . . . | 42.25 |
| Silica . . . . . . . | 23.00 |
| Alumina and oxide of iron . | 31.00 |
| Phosphate of lime . . . . | 0.05 |
| | 99.30 |

* Walchner's *Geognosie* (German), p. 387.

† Prof. M. Tuomey's *Report*, p. 135.

ANALYSIS OF ASH OF PINE WOOD.*

| | |
|---|---|
| Silica . . . . . | 10.8667 |
| Sulphuric acid . . . | 1.2844 |
| Phosphoric acid . . . | 3.5569 |
| Chlorine . . . . | 0.1229 |
| Peroxide of iron . . . | 2.6018 |
| Protoxide of manganese . | 2.6498 |
| Magnesia . . . . . | 3.9873 |
| Lime . . . . . . | 58.6475 |
| Potash . . . . . | 2.3076 |
| Soda . . . . . . | 13.9751 |

* This analysis is by Sacc (*Annales de Chimie et de Physique*, xxv.), given also in Booth and Morfit's *Smithsonian Report*, p. 201. It should have been placed in the Article on Ashes of Plants, &c., p. 242, but was then omitted.

## ANALYSIS OF WATER FROM THE DEAD SEA, BY GMELIN.

| | |
|---|---|
| Chloride of lime . . | 3.21410 |
| Chloride of magnesium . | 11.31724 |
| Chloride of sodium . . | 7.07617 |
| Chloride of potassium . | 1.67000 |
| Chloride of aluminium . | 0.89600 |
| Chloride of manganese . | 0.21170 |
| Chloride of ammonium . | 0.00715 |
| Bromide of magnesium . | 0.43970 |
| Sulphate of lime . . | 0.05270 |
| Total of saline ingredients | 24.5398 |
| Water . . . . . | 75.4602 |
| | 100.0000 |

## TABLE OF THE COMBINING PROPORTIONS OF THE ELEMENTS, OXYGEN BEING EQUAL TO 100.

| Element. | Symbol. | | Equiv. |
|---|---|---|---|
| Oxygen | O | ... | 100.00 |
| Sulphur | S | ... | 200.75 |
| Selenium | Se | ... | 495.28 |
| Tellurium | Te | ... | 801.76 |
| | | | |
| Nitrogen | N | ... | 175.06 |
| Phosphorus | P | ... | 392.04 |
| Arsenic | As | ... | 938.80 |
| Antimony | Sb | ... | 1612.90 |
| | | | |
| Chlorine | Cl | ... | 443.28 |
| Bromine | Br | ... | 999.62 |
| Iodine | I | ... | 1585.99 |
| Fluorine | F | ... | 235.48 |
| Carbon | C | ... | 75.12 |
| Boron | B | ... | 136.20 |
| Silicon | Si | ... | 277.78 |
| | | | |
| Potassium (Kalium) | K | ... | 488.85 |
| Soda (Natrium) | Na | ... | 289.73 |
| Lithium | L | ... | 81.66 |
| Barium | Ba | ... | 855.29 |
| Strontium | Sr | ... | 545.93 |
| Calcium | Ca | ... | 251.65 |
| Magnesium | Mg | ... | 158.14 |

| Element. | Symbol. | | Equiv. |
|---|---|---|---|
| Aluminium | Al | ... | 170.90 |
| Beryllium (Glucinium) | G | ... | 87.12 |
| Yttrium | Y | ... | 402.51 |
| Terbium | T | ... | ? |
| Erbium | E | ... | ? |
| Zirconium | Z or Zr | ... | 420.20 |
| Norium | No | ... | ? |
| Thorium | Th | ... | 744.86 |
| Cerium | Ce | ... | 574.7 |
| Lanthanium | La | ... | ? |
| Didymium | D | ... | ? |
| | | | |
| Iron | Fe | ... | 350.53 |
| Manganese | Mn | ... | 344.68 |
| Cobalt | Co | ... | 368.65 |
| Nickel | Ni | ... | 369.33 |
| Zinc | Zn | ... | 406.59 |
| Cadmium | Cd | ... | 696.77 |
| Tin (Stannum) | Sn | ... | 735.29 |
| Uranium | U | ... | 742.87 |
| | | | |
| Lead (Plumbum) | Pb | ... | 1294.64 |
| Bismuth | Bi | ... | 1330.37 |
| Copper (Cuprum) | Cu | ... | 395.60 |
| Mercury (Hydrargyrum) | Hg | ... | 1251.29 |
| Silver (Argentium) | Ag | ... | 1349.66 |
| Palladium | Pd | ... | 665.47 |
| Ruthenium | Ru | ... | 646.27 |
| Rhodium | R | ... | 651.96 |

| Element. | Symbol. | | Equiv. |
|---|---|---|---|
| PLATINUM | Pt | ... | 1232.08 |
| IRIDIUM | Ir | ... | 1232.08 |
| GOLD (AURUM) | Au | ... | 1229.16 |
| OSMIUM | Os | ... | 1242.62 |
| | | | |
| TITANIUM | Ti | ... | 301.55 |
| TANTALUM | Ta | ... | ? |
| NIOBIUM | Nb | ... | ? |
| PELOPIUM | Pp | ... | ? |
| | | | |
| TUNGSTEN (WOLFRAM) | W | ... | 1188.36 |
| MOLYBDENUM | Mo | ... | 575.83 |
| VANADIUM | V | ... | 856.89 |
| CHROMIUM | Cr | ... | 328.87 |
| | | | |
| HYDROGEN | H | ... | 12.48 |

# GLOSSARY.

*Acidulate, to,* is to mix with free acid.

*Alkaline,* containing free alkali.

*Alloy,* a mixture of reduced metals.

*Amalgam,* an alloy in which mercury forms a part.

*Amorphous,* syn. uncrystalline, without form.

*Anhydrous,* possessing no water.

*Atmosphere* of a gas, signifies a surrounding volume of gas; often syn. current.

*Behavior* of a substance, is the manner in which it acts in any peculiar situation.

*Botryoidal,* shaped like bunches of grapes, from the Greek βοτρυωδης.

*Calcine,* to heat to redness, originally only of lime.

*Caustic;* an alkali, for instance, is caustic if uncombined.

*Chyme,* from the Greek, is the grayish pulp formed from the food after it has remained in the stomach for some time.

*Comportment,* syn. behavior.

*Capillation,* a peculiar manner of treating silver ores in assaying them by heat.

*Decrepitate, to.* A salt decrepitates, if on being exposed to a red heat the water is discharged, and the substance loses its crystalline form.

*Deliquescent.* A substance is deliquescent, if on exposure to the action of the atmospheric air, it absorbs a sufficient quantity of hygroscopic moisture to become fluid.

*Digest, to.* This term denotes the submitting a liquid or solid to the action of some liquid, with application of a moderate (digestive) heat (*see* INTRODUCTION).

*Ebullition,* is the phenomenon produced when some gas or vapor is disengaged from a liquid with force, as in the escaping of steam from water.

*Effervescence* is a similar phenomenon, being also the escaping of gas, but with the production of froth or foam. Acid on carbonate of lime produces effervescence.

*Evolve, to.* As to discharge a gas, or to cause it to be discharged.

*Excess.* An excess of an acid, alkali, salt, or other substance signifies, chemically speaking, a quantity over and above that which will saturate or neutralize another substance, with which it is brought in contact.

*Filtrate.* A filtrate is the liquid which is filtered off from a precipitate.

*Fulminate, to,* is to burn with more or less detonation.

*Fuming acids* are those which on exposure to the atmosphere without heat produce vapors. They either contain some volatile ingredients, as fuming nitric acid, or are of such strength that the evaporating particles absorb atmospheric moisture, and thus produce vapors, as with the Nordhausen oil of vitriol.

*Gangue*, is the rock, a dike or vein, which contains some mineral; derived from the German: *gang*.

*Humic compounds* are produced from vegetable mould, viz. the humus or humine and the humic acid. The term is not very definite.

*Hydrate.* A substance is termed a hydrate if it possess any considerable amount of water.

*Hygroscopic*, from the Greek, is the name applied to the moisture, absorbed by substances from the atmospheric air, in contradistinction to the water of crystallization.

*Leach, to.* See INTRODUCTION.

*Macerate, to*, a substance, generally signifies to submit it to the action of a liquid, in which only a portion is soluble, but always means to expose every portion to its influence.

*Matrix*, of a mineral or metal, is the rock in which it occurs. It often is synonymous to gangue, though the term matrix is also applied to massive metalliferous rocks.

*Metals*, in chemical terminology are not only what in common parlance are termed thus. They all possess the peculiar appearance termed metallic lustre, are neither transparent nor translucent, and are all capable of being molten. Many of them are volatile, and many occur in a crystalline form, or can be artificially caused to crystallize. The cube is the base of their crystallization. Many metals are not really known as such, and are only at present imaginary, as for example silicon, of which only the oxide, silica or silicic acid is known.

*Neutralize, to,* a solution, is to balance the amount of alkali and acid in it.

*Oxidize, to,* is to convert into an oxide.

*Precipitate, to,* signifies to cause the formation of a deposit in a liquid. A precipitate is, therefore, the deposit produced.

*Quartation,* or quartering, is the name applied to a part of the process employed in assaying gold ores by heat.

*Reduce, to,* is to diminish the quantity of oxygen in an oxide. Generally, it means the converting of an oxide into the metallic state, though it may also signify to alter an oxide possessing much oxygen into one containing less, as to reduce a peroxide to a protoxide.

*Roast, to;* syn. calcine.

*Salts,* are the compounds of metals with simple or compound salt radicals.

*Saturate, to.* A substance is saturated with another, if the utmost quantity of the latter is added to it that the former is capable of combining with, without a free portion of either being left.

*Seething,* temperature. A solution is said to be at a seething temperature, when it is almost entering upon the boiling state. It is not a very definite term. The temperature required to place water in this state might, under ordinary circumstances, perhaps, be considered to be about 200° Fahr.

*Solution.* A solution is a liquid in which some substance is chemically suspended.

*Sublimate.* A sublimate is a substance which has been in a state of vapor, and is then again condensed, reduced to its solid form.

*Surplus*, syn. excess.

*Throw down, to;* syn. to precipitate.

*Wash, to.* See description in the INTRODUCTION.

*Washings*, are those liquids which have been employed for washing, and have been permitted to percolate through the filter, and the substance on the filter, and which contains more or less of the contents of the first filtrate, but always in a very diluted form.

*Water of crystallization*, is that water which is required by a substance for the formation of its crystals, and by removing which it becomes amorphous.

# INDEX.

THE END.

# PUBLICATIONS

OF

# HENRY CAREY BAIRD,

SUCCESSOR TO E. L. CAREY,

S. E. CORNER OF MARKET AND FIFTH STREETS,

PHILADELPHIA.

## SCIENTIFIC AND PRACTICAL.

### THE PRACTICAL MODEL CALCULATOR,

For the Engineer, Machinist, Manufacturer of Engine Work, Naval Architect, Miner, and Millwright. By Oliver Byrne, Compiler and Editor of the Dictionary of Machines, Mechanics, Engine Work and Engineering, and Author of various Mathematical and Mechanical Works. Illustrated by numerous Engravings. To be published in Twelve Parts, at Twenty-five Cents each, forming, when completod, One large Volume, Octavo, of nearly six hundred pages.

It will contain such calculations as are met with and required in the Mechanical Arts, and establish models or standards to guide practical men. The Tables that are introduced, many of which are new, will greatly economize labour, and render the every-day calculations of the *practical man* comprehensive and easy. From every single calculation given in this work numerous other calculations are readily modelled, so that each may be considered the head of a numerous family of practical results.

The examples selected will be found appropriate, and in all cases taken from the actual practice of the present time. Every rule has been tested by the unerring results of mathematical research, and confirmed by experiment, when such was necessary.

The Practical Model Calculator will be found to fill a vacancy in the library of the practical working-man long considered a requirement. It will be found to excel all other works of a similar nature, from the great extent of its range, the exemplary nature of its well-selected examples, and from the easy, simple, and systematic manner in which the model calculations are established.

### NORRIS'S HAND-BOOK FOR LOCOMOTIVE ENGINEERS AND MACHINISTS:

Comprising the Calculations for Constructing Locomotives. Manner of setting Valves, &c. &c. By Septimus Norris, Civil and Mechanical Engineer. In One Volume, 12mo., with illustrations........................................................ $1.50

## A COMPLETE TREATISE ON TANNING, CURRYING AND EVERY BRANCH OF LEATHER-DRESSING.

From the French and from original sources. By CAMPBELL MORFIT, one of the Editors of the "Encyclopedia of Chemistry," Author of "Chemistry Applied to the Manufacture of Soap and Candles," and other Scientific Treatises. Illustrated with several hundred Engravings. Complete in One Vol., royal 8vo. (In press.)

This important treatise will be issued from the press at as early a day as the duties of the editor will permit, and it is believed that in no other branch of applied science could more signal service be rendered to American Manufacturers.

The publisher is not aware that in any other work heretofore issued in this country, more space has been devoted to this subject than a single chapter; and in offering this volume to so large and intelligent a class as American Tanners and Leather Dressers, he feels confident of their substantial support and encouragement.

---

## THE PRACTICAL COTTON-SPINNER AND MANUFACTURER; Or, The Manager's and Overseer's Companion.

This works contains a Comprehensive System of Calculations for Mill Gearing and Machinery, from the first moving power through the different processes of Carding, Drawing, Slabbing, Roving, Spinning, and Weaving, adapted to American Machinery, Practice, and Usages. Compendious Tables of Yarns and Reeds are added. Illustrated by large Working-drawings of the most approved American Cotton Machinery. Complete in One Volume, octavo..............................................................$3.50

This edition of Scott's Cotton-Spinner, by OLIVER BYRNE, is designed for the American Operative. It will be found intensely practical, and will be of the greatest possible value to the Manager, Overseer, and Workman.

---

## THE PRACTICAL METAL-WORKER'S ASSISTANT,

For Tin-Plate Workers, Brasiers, Coppersmiths, Zinc-Plate Ornamenters and Workers, Wire Workers, Whitesmiths, Blacksmiths, Bell Hangers, Jewellers, Silver and Gold Smiths, Electrotypers, and all other Workers in Alloys and Metals. By CHARLES HOLTZAPPFEL. Edited, with important additions, by OLIVER BYRNE. Complete in One Volume, octavo............$4.00

It will treat of Casting, Founding, and Forging; of Tongs and other Tools; Degrees of Heat and Managemnet of Fires; Welding; of Heading and Swage Tools; of Punches and Anvils; of Hardening and Tempering; of Malleable Iron Castings, Case Hardening, Wrought and Cast Iron. The management and manipulation of Metals and Alloys, Melting and Mixing. The management of Furnaces, Casting and Founding with Metallic Moulds, Joining and Working Sheet Metal. Peculiarities of the different Tools employed. Processes dependent on the ductility of Metals. Wire Drawing, Drawing Metal Tubes, Soldering. The use of the Blowpipe, and every other known Metal-Worker's Tool. To the works of Holtzappfel, OLIVER BYRNE has added all that is useful and peculiar to the American Metal-Worker.

## THE MANUFACTURE OF IRON IN ALL ITS VARIOUS BRANCHES:

To which is added an Essay on the Manufacture of Steel, by FREDERICK OVERMAN, Mining Engineer, with one hundred and fifty Wood Engravings. A new edition. In One Volume, octavo, five hundred pages........................................$5.00

We have now to announce the appearance of another valuable work on the subject which, in our humble opinion, supplies any deficiency which late improvements and discoveries may have caused, from the lapse of time since the date of "Mushet" and "Schrivenor." It is the production of one of our transatlantic brethren, Mr. Frederick Overman, Mining Engineer; and we do not hesitate to set it down as a work of great importance to all connected with the iron interest; one which, while it is sufficiently technological fully to explain chemical analysis, and the various phenomena of iron under different circumstances, to the satisfaction of the most fastidious, is written in that clear and comprehensive style as to be available to the capacity of the humblest mind, and consequently will be of much advantage to those works where the proprietors may see the desirability of placing it in the hands of their operatives.—*London Morning Journal.*

---

## A TREATISE ON THE AMERICAN STEAM-ENGINE.

Illustrated by numerous Wood Cuts and other Engravings. By OLIVER BYRNE. In One Volume, royal 8vo. (In press.)

---

## PROPELLERS AND STEAM NAVIGATION:

With Biographical Sketches of Early Inventors. By ROBERT MACFARLANE, C. E., Editor of the "Scientific American." In One Volume, 12mo. Illustrated by over Eighty Wood Engravings..............................................................75 cts.

The object of this "History of Propellers and Steam Navigation" is twofold. One is the arrangement and description of many devices which have been invented to propel vessels, in order to prevent many ingenious men from wasting their time, talents, and money on such projects. The immense amount of time, study, and money thrown away on such contrivances is beyond calculation. In this respect, it is hoped that it will be the means of doing some good.—*Preface.*

---

## MATHEMATICS FOR PRACTICAL MEN:

Being a Common-place Book of Principles, Theorems, Rules, and tables, in various Departments of Pure and Mixed Mathematics, with their applications, especially to the pursuits of Surveyors, Architects, Mechanics, and Civil Engineers. With numerous Engravings. By OLINTHUS GREGORY, LL.D., F. R. A. S..............................................................$1.50

Only let men awake, and fix their eye, one while on the nature of things, another while on the application of them to the use and service of mankind.—*Lord Bacon.*

## PRACTICAL SERIES.

THE volumes in this series are published in duodecimo form, and the design is to furnish to Artisans, for a moderate sum, Hand-books of the different Arts and Manufactures, in order that they may be enabled to keep pace with the improvements of the age.

There have already appeared—

THE AMERICAN MILLER AND MILLWRIGHT'S ASSISTANT. $1.
THE TURNER'S COMPANION. 75 cts.
THE PAINTER, GILDER, AND VARNISHER'S COMPANION. 75 cts.
THE DYER AND COLOUR-MAKER'S COMPANION. 75 cts.
THE BUILDER'S COMPANION. $1.
THE CABINET-MAKER'S COMPANION. 75 cts.

The following, among others, are in preparation:—

A TREATISE ON A BOX OF INSTRUMENTS. By THOMAS KENTISH.
THE PAPER-HANGER'S COMPANION. By J. ARROWSMITH.

## THE AMERICAN MILLER AND MILLWRIGHT'S ASSISTANT:

By WILLIAM CARTER HUGHES, Editor of "The American Miller," (newspaper,) Buffalo, N. Y. Illustrated by Drawings of the most approved Machinery. In One Volume, 12mo.........$1

The author offers it as a substantial reference, instead of speculative theories, which belong only to those not immediately attached to the business. Special notice is also given of most of the essential improvements which have of late been introduced for the benefit of the Miller.—*Savannah Republican.*

The whole business of making flour is most thoroughly treated by him.—*Bulletin.*

A very comprehensive view of the Millwright's business.—*Southern Literary Messenger.*

## THE TURNER'S COMPANION:

Containing Instructions in Concentric, Elliptic, and Eccentric Turning. Also, various Plates of Chucks, Tools, and Instruments, and Directions for using the Eccentric Cutter, Drill, Vertical Cutter, and Circular Rest; with Patterns and Instructions for working them. Illustrated by numerous Engravings. In One Volume, 12mo..........................................75 cts.

The object of the Turner's Companion is to explain in a clear, concise, and intelligible manner, the rudiments of this beautiful art.—*Savannah Republican.*

There is no description of turning or lathe-work that this elegant little treatise does not describe and illustrate.—*Western Lit. Messenger.*

## THE PAINTER, GILDER, AND VARNISHER'S COMPANION:

Containing Rules and Regulations for every thing relating to the arts of Painting, Gilding, Varnishing, and Glass Staining; numerous useful and valuable Receipts; Tests for the detection of Adulterations in Oils, Colours, &c., and a Statement of the Diseases and Accidents to which Painters, Gilders, and Varnishers are particularly liable; with the simplest methods of Prevention and Remedy. In one vol. small 12mo., cloth. 75cts.

Rejecting all that appeared foreign to the subject, the compiler has omitted nothing of real practical worth.—*Hunt's Merchant's Magazine.*

An excellent *practical work*, and one which the practical man cannot afford to be without.—*Farmer and Mechanic.*

It contains every thing that is of interest to persons engaged in this trade. —*Bulletin.*

This book will prove valuable to all whose business is in any way connected with painting.—*Scott's Weekly.*

Cannot fail to be useful.—*N. Y. Commercial.*

---

## THE BUILDER'S POCKET COMPANION:

Containing the Elements of Building, Surveying, and Architecture; with Practical Rules and Instructions connected with the subject. By A. C. Smeaton, Civil Engineer, &c. In one volume, 12mo. $1.

Contents:—The Builder, Carpenter, Joiner, Mason, Plasterer, Plumber, Painter, Smith, Practical Geometry, Surveyor, Cohesive Strength of Bodies, Architect.

It gives, in a small space, the most thorough directions to the builder, from the laying of a brick, or the felling of a tree, up to the most elaborate production of ornamental architecture. It is scientific, without being obscure and unintelligible, and every house-carpenter, master, journeyman, or apprentice, should have a copy at hand always.—*Evening Bulletin.*

Complete on the subjects of which it treats. A most useful practical work. —*Balt. American.*

It must be of great practical utility.—*Savannah Republican.*

To whatever branch of the art of building the reader may belong, he will find in this something valuable and calculated to assist his progress.—*Farmer and Mechanic.*

This is a valuable little volume, designed to assist the student in the acquisition of elementary knowledge, and will be found highly advantageous to every young man who has devoted himself to the interesting pursuits of which it treats.—*Va. Herald.*

## THE DYER AND COLOUR-MAKER'S COMPANION:

Containing upwards of two hundred Receipts for making Colors, on the most approved principles, for all the various styles and fabrics now in existence; with the Scouring Process, and plain Directions for Preparing, Washing-off, and Finishing the Goods. In one volume, small 12mo., cloth. 75 cts.

This is another of that most excellent class of practical books, which the publisher is giving to the public. Indeed we believe there is not, for manufacturers, a more valuable work, having been prepared for, and expressly adapted to their business.—*Farmer and Mechanic.*

It is a valuable book.—*Otsego Republican.*

We have shown it to some practical men, who all pronounced it the completest thing of the kind they had seen—*N. Y. Nation.*

---

## THE CABINET-MAKER AND UPHOLSTERER'S COMPANION:

Comprising the Rudiments and Principles of Cabinet Making and Upholstery, with familiar instructions, illustrated by Examples, for attaining a proficiency in the Art of Drawing, as applicable to Cabinet Work; the processes of Veneering, Inlaying, and Buhl Work; the art of Dyeing and Staining Wood, Ivory, Bone, Tortoise-shell, etc. Directions for Lackering, Japanning, and Varnishing; to make French Polish; to prepare the best Glues, Cements, and Compositions, and a number of Receipts particularly useful for Workmen generally, with Explanatory and Illustrative Engravings. By J. Stokes. In one volume, 12mo., with illustrations. 75 cts.

---

## THE PAPER-HANGER'S COMPANION:

In which the Practical Operations of the Trade are systematically laid down; with copious Directions Preparatory to Papering; Preventions against the effect of Damp in Walls; the various Cements and Pastes adapted to the several purposes of the Trade; Observations and Directions for the Panelling and Ornamenting of Rooms, &c. &c. By James Arrowsmith. In One Volume, 12mo. (In press.)

## THE FRUIT, FLOWER, AND KITCHEN GARDEN.

By Patrick Neill, L.L.D.

Thoroughly revised, and adapted to the climate and seasons of the United States, by a Practical Horticulturist. Illustrated by numerous Engravings. In one volume, 12mo. $1.25.

---

## HOUSEHOLD SURGERY; OR, HINTS ON EMERGENCIES.

By J. F. South, one of the Surgeons of St. Thomas's Hospital. In one volume, 12mo. Illustrated by nearly fifty Engravings. $1.25.

CONTENTS:

*The Doctor's Shop.*—Poultices, Fomentations, Lotions, Liniments, Ointments, Plasters.

*Surgery.*—Blood-letting, Blistering, Vaccination, Tooth-drawing, How to put on a Roller, Lancing the Gums, Swollen Veins, Bruises, Wounds, Torn or Cut Achilles Tendon, What is to be done in cases of sudden Bleeding from various causes, Scalds and Burns, Frost-bite, Chilblains, Sprains, Broken Bones, Bent Bones, Dislocations, Ruptures, Piles, Protruding Bowels, Wetting the Bed, Whitlow, Boils, Black-heads, Ingrowing Nails, Bunions, Corns, Sty in the Eye, Blight in the Eye, Tumours in the Eyelids, Inflammation on the Surface of the Eye, Pustules on the Eye, Milk Abscesses, Sore Nipples, Irritable Breast, Breathing, Stifling, Choking, Things in the Eye, On Dress, Exercise and Diet of Children, Bathing, Infections, Observations on Ventilation.

---

## HOUSEHOLD MEDICINE.

In one volume, 12mo. Uniform with, and a companion to, the above. (In immediate preparation.)

## THE ENCYCLOPEDIA OF CHEMISTRY, PRACTICAL AND THEORETICAL:

Embracing its application to the Arts, Metallurgy, Mineralogy, Geology, Medicine, and Pharmacy. By JAMES C. BOOTH, Melter and Refiner in the United States Mint; Professor of Applied Chemistry in the Franklin Institute, etc.; assisted by CAMPBELL MORFIT, author of "Chemical Manipulations," etc. Complete in one volume, royal octavo, 978 pages, with numerous wood cuts and other illustrations. $5.

It covers the whole field of Chemistry as applied to Arts and Sciences. * * * As no library is complete without a common dictionary, it is also our opinion that none can be without this Encyclopedia of Chemistry.—*Scientific American.*

A work of time and labour, and a treasury of chemical information.—*North American.*

By far the best manual of the kind which has been presented to the American public.—*Boston Courier.*

An invaluable work for the dissemination of sound practical knowledge.—*Ledger.*

A treasury of chemical information, including all the latest and most important discoveries.—*Baltimore American.*

At the first glance at this massive volume, one is amazed at the amount of reading furnished in its compact double pages, about one thousand in number. A further examination shows that every page is richly stored with information, and that while the labours of the authors have covered a wide field, they have neglected or slighted nothing. Every chemical term, substance, and process is elaborately, but intelligibly, described. The whole science of Chemistry is placed before the reader as fully as is practicable, with a science continually progressing. * * Unlike most American works of this class, the authors have not depended upon any one European work for their materials. They have gathered theirs from works on Chemistry in all languages, and in all parts of Europe and America; their own experience, as practical chemists, being ever ready to settle doubts or reconcile conflicting authorities. The fruit of so much toil is a work that must ever be an honour to American Science.—*Evening Bulletin.*

---

## PERFUMERY; ITS MANUFACTURE AND USE:

With Instructions in every branch of the Art, and Receipts for all the Fashionable Preparations; the whole forming a valuable aid to the Perfumer, Druggist, and Soap Manufacturer. Illustrated by numerous Wood-cuts. From the French of Celnart, and other late authorities. With Additions and Improvements by CAMPBELL MORFIT, one of the Editors of the "Encyclopedia of Chemistry." In one volume, 12mo., cloth. $1.

## A TREATISE ON A BOX OF INSTRUMENTS,

And the Slide Rule, with the Theory of Trigonometry and Logarithms, including Practical Geometry, Surveying, Measuring of Timber, Cask and Malt Gauging, Heights and Distances. By Thomas Kentish. In One Volume, 12mo. (In press.)

---

## STEAM FOR THE MILLION.

An Elementary Outline Treatise on the Nature and Management of Steam, and the Principles and Arrangement of the Engine. Adapted for Popular Instruction, for Apprentices, and for the use of the Navigator. With an Appendix containing Notes on Expansive Steam, &c. In One Volume, 8vo...37½ cts.

---

## SYLLABUS OF A COMPLETE COURSE OF LECTURES ON CHEMISTRY:

Including its Application to the Arts, Agriculture, and Mining, prepared for the use of the Gentlemen Cadets at the Hon. E. I. Co.'s Military Seminary, Addiscombe. By Professor E. Solly, Lecturer on Chemistry in the Hon. E. I. Co.'s Military Seminary. Revised by the Author of "Chemical Manipulations." In one volume, octavo, cloth. $1.25.

The present work is designed to occupy a vacant place in the libraries of Chemical text-books. It is admirably adapted to the wants of both teacher and pupil; and will be found especially convenient to the latter, either as a companion in the class-room, or as a remembrancer in the study. It gives, at a glance, under appropriate headings, a classified view of the whole science, which is at the same time compendious and minutely accurate; and its wide margins afford sufficient blank space for such manuscript notes as the student may wish to add during lectures or recitations.

The almost indispensable advantages of such an impressive aid to memory are evident to every student who has used one in other branches of study. Therefore, as there is now no Chemical Syllabus, we have been induced by the excellencies of this work to recommend its republication in this country; confident that an examination of the contents will produce full conviction of its intrinsic worth and usefulness.—*Editor's Preface.*

## ELECTROTYPE MANIPULATION:

Being the Theory and Plain Instructions in the Art of Working in Metals, by Precipitating them from their Solutions, through the agency of Galvanic or Voltaic Electricity. By Charles V. Walker, Hon. Secretary to the London Electrical Society, etc. Illustrated by Wood-cuts. In one volume, 24mo., cloth. From the thirteenth London edition. 62 cts.

---

## PHOTOGENIC MANIPULATION:

Containing the Theory and Plain Instructions in the Art of Photography, or the Productions of Pictures through the Agency of Light; including Calotype, Chrysotype, Cyanotype, Chromatype, Energiatype, Anthotype, Amphitype, Daguerreotype, Thermography, Electrical and Galvanic Impressions. By George Thomas Fisher, Jr., Assistant in the Laboratory of the London Institution. Illustrated by wood-cuts. In one volume, 24mo., cloth. 62 cts.

---

## MATHEMATICS FOR PRACTICAL MEN:

Being a Common-Place Book of Principles, Theorems, Rules, and Tables, in various departments of Pure and Mixed Mathematics, with their Applications; especially to the pursuits of Surveyors, Architects, Mechanics, and Civil Engineers, with numerous Engravings. By Olinthus Gregory, L. L. D. $1.50.

Only let men awake, and fix their eyes, one while on the nature of things, another while on the application of them to the use and service of mankind. —*Lord Bacon.*

---

## AN ELEMENTARY COURSE OF INSTRUCTION ON ORDNANCE AND GUNNERY:

Prepared for the use of the Midshipmen at the Naval School. By James H. Ward, U. S. N. In one volume, octavo. $1.50.

## SHEEP-HUSBANDRY IN THE SOUTH:

Comprising a Treatise on the Acclimation of Sheep in the Southern States, and an Account of the different Breeds. Also, a Complete Manual of Breeding, Summer and Winter Management, and of the Treatment of Diseases. With Portraits and other Illustrations. By Henry S. Randall. In One Volume, octavo..........................................................................$1.25

---

## ELWOOD'S GRAIN TABLES:

Showing the value of Bushels and Pounds of different kinds of Grain, calculated in Federal Money, so arranged as to exhibit upon a single page the value at a given price from *ten cents to two dollars* per bushel, of any quantity from *one pound to ten thousand bushels.* By J. L. Elwood. A new Edition. In One Volume, 12mo..........................................................................$1

To Millers and Produce Dealers this work is pronounced by all who have it in use, to be superior in arrangement to any work of the kind published—and *unerring accuracy in every calculation may be relied upon in every instance.*

☞ A reward of Twenty-five Dollars is offered for an error of one cent found in the work.

---

## MISS LESLIE'S COMPLETE COOKERY.

Directions for Cookery, in its Various Branches. By Miss Leslie. Forty-second Edition. Thoroughly Revised, with the Addition of New Receipts. In One Volume, 12mo, half bound, or in sheep..........................................................................$1

In preparing a new and carefully revised edition of this my first work on cookery, I have introduced improvements, corrected errors, and added new receipts, that I trust will on trial be found satisfactory. The success of the book (proved by its immense and increasing circulation) affords conclusive evidence that it has obtained the approbation of a large number of my country-women; many of whom have informed me that it has made practical house-wives of young ladies who have entered into married life with no other acquirements than a few showy accomplishments. Gentlemen, also, have told me of great improvements in the family table, after presenting their wives with this manual of domestic cookery, and that, after a morning devoted to the fatigues of business, they no longer find themselves subjected to the annoyance of an ill-dressed dinner.—*Preface.*

---

## MISS LESLIE'S TWO HUNDRED RECEIPTS IN FRENCH COOKERY.

A new Edition, in cloth..........................................25 cts.

## TABLES OF LOGARITHMS FOR ENGINEERS AND MACHINISTS:

Containing the Logarithms of the Natural Numbers, from 1 to ₁00000, by the help of Proportional Differences. And Logarithmic Sines, Cosines, Tangents, Co-tangents, Secants, and Co-secants, for every Degree and Minute in the Quadrant. To which are added, Differences for every 100 Seconds. By Oliver Byrne, Civil, Military, and Mechanical Engineer. In One Volume, 8vo. cloth............................................$1

---

## TWO HUNDRED DESIGNS FOR COTTAGES AND VILLAS, &c. &c.

Original and Selected. By Thomas U. Walter, Architect of Girard College, and John Jay Smith, Librarian of the Philadelphia Library. In Four Parts, quarto.......................$10

---

## ELEMENTARY PRINCIPLES OF CARPENTRY.

By Thomas Tredgold. In One Volume, quarto, with numerous Illustrations............................................$2.50

---

## A TREATISE ON BREWING AND DISTILLING,

In One Volume, 8vo. (In press.)

---

## FAMILY ENCYCLOPEDIA

Of Useful Knowledge and General Literature; containing about Four Thousand Articles upon Scientific and Popular Subjects. With Plates. By John L. Blake, D. D. In One Volume, 8vo, full bound............................................$5

---

## SYSTEMATIC ARRANGEMENT OF COKE'S FIRST INSTITUTES OF THE LAWS OF ENGLAND.

By J. H. Thomas. Three Volumes, 8vo, law sheep.........$12

---

## AN ACCOUNT OF SOME OF THE MOST IMPORTANT DISEASES OF WOMEN.

By Robert Gooch, M. D. In One Volume, 8vo, sheep...$1.50

# STANDARD ILLUSTRATED POETRY.

## THE TALES AND POEMS OF LORD BYRON:

Illustrated by Henry Warren. In One Volume, royal 8vo. with 10 Plates, scarlet cloth, gilt edges..............................$5
Morocco extra..............................................................$7

It is illustrated by several elegant engravings, from original designs by Warren, and is a most splendid work for the parlour or study.—*Boston Evening Gazette.*

## CHILDE HAROLD; A ROMAUNT BY LORD BYRON:

Illustrated by 12 Splendid Plates, by Warren and others. In One Volume, royal 8vo., cloth extra, gilt edges..................$5
Morocco extra .............................................................$7

Printed in elegant style, with splendid pictures, far superior to any thing of the sort usually found in books of this kind.—*N. Y. Courier.*

## THE FEMALE POETS OF AMERICA.

By Rufus W. Griswold. A new Edition. In One Volume, royal 8vo. Cloth, gilt.........................................$2.50
Cloth extra, gilt edges.....................................................$3
Morocco super extra ....................................................$4.50

The best production which has yet come from the pen of Dr. Griswold, and the most valuable contribution which he has ever made to the literary celebrity of the country.—*N. Y. Tribune.*

## THE LADY OF THE LAKE:

By Sir Walter Scott. Illustrated with 10 Plates, by Corbould and Meadows. In One Volume, royal 8vo. Bound in cloth extra, gilt edges...............................................$5
Turkey morocco super extra............................................$7

This is one of the most truly beautiful books which has ever issued from the American press.

## LALLA ROOKH; A ROMANCE BY THOMAS MOORE:

Illustrated by 13 Plates, from Designs by Corbould, Meadows, and Stephanoff. In One Volume, royal 8vo. Bound in cloth extra, gilt edges...............................................$5
Turkey morocco super extra............................................$7

This is published in a style uniform with the "Lady of the Lake."

## THE POETICAL WORKS OF THOMAS GRAY:

With Illustrations by C. W. Radcliff. Edited with a Memoir, by Henry Reed, Professor of English Literature in the University of Pennsylvania. In One Volume, 8vo. Bound in cloth extra, gilt edges..........$3.50
Turkey morocco super extra..........$5.50

It is many a day since we have seen issued from the press of our country a volume so complete and truly elegant in every respect. The typography is faultless, the illustrations superior, and the binding superb.—*Troy Whig.*

We have not seen a specimen of typographical luxury from the American press which can surpass this volume in choice elegance.—*Boston Courier.*

It is eminently calculated to consecrate among American readers, (if they have not been consecrated already in their hearts,) the pure, the elegant, the refined, and, in many respects, the sublime imaginings of Thomas Gray.—*Richmond Whig.*

---

## THE POETICAL WORKS OF HENRY WADSWORTH LONGFELLOW:

Illustrated by 10 Plates, after Designs by D. Huntingdon, with a Portrait. Ninth Edition. In One Volume, royal 8vo. Bound in cloth extra, gilt edges..........$5
Morocco super extra..........$7

This is the very luxury of literature—Longfellow's charming poems presented in a form of unsurpassed beauty.—*Neal's Gazette.*

---

## POETS AND POETRY OF ENGLAND IN THE NINETEENTH CENTURY.

By Rufus W. Griswold. Illustrated. In One Volume, royal 8vo. Bound in cloth..........$3
Cloth extra, gilt edges..........$3.50
Morocco super extra..........$5

Such is the critical acumen discovered in these selections, that scarcely a page is to be found but is redolent with beauties, and the volume itself may be regarded as a galaxy of literary pearls.—*Democratic Review.*

---

## THE TASK, AND OTHER POEMS.

By William Cowper. Illustrated by 10 Steel Engravings. In One Volume, 12mo. Cloth extra, gilt edges..........$2
Morocco extra..........$2

"The illustrations in this edition of Cowper are most exquisitely designed and engraved."

## THE FEMALE POETS OF GREAT BRITAIN.

With Copious Selections and Critical Remarks. By Frederic Rowton. With Additions by an American Editor, and finely engraved Illustrations by celebrated Artists. In One Volume, royal 8vo. Bound in cloth extra, gilt edges....................$5
Turkey morocco..................................................$7

Mr. Rowton has presented us with admirably selected specimens of nearly one hundred of the most celebrated female poets of Great Britain, from the time of Lady Juliana Bernes, the first of whom there is any record, to the Mitfords, the Hewitts, the Cooks, the Barretts, and others of the present day.—*Hunt's Merchants' Magazine.*

---

## SPECIMENS OF THE BRITISH POETS.

From the time of Chaucer to the end of the Eighteenth Century. By Thomas Campbell. In One Volume, royal 8vo. (In press.)

---

## THE POETS AND POETRY OF THE ANCIENTS:

By William Peter, A. M. Comprising Translations and Specimens of the Poets of Greece and Rome, with an elegant engraved View of the Coliseum at Rome. Bound in cloth......$3
Cloth extra, gilt edges..........................................$3.50
Turkey morocco super extra......................................$5

It is without fear that we say that no such excellent or complete collection has ever been made. It is made with skill, taste, and judgment.—*Charleston Patriot.*

---

## THE POETICAL WORKS OF N. PARKER WILLIS.

Illustrated by 16 Plates, after designs by E. Leutze. In One Volume, royal 8vo. A new Edition. Bound in cloth extra, gilt edges..........................................................$5
Turkey morocco super extra......................................$7

This is one of the most beautiful works ever published in this country.—*Courier and Inquirer.*

Pure and perfect in sentiment, often in expression, and many a heart has been won from sorrow or roused from apathy by his earlier melodies. The illustrations are by Leutze,—a sufficient guarantee for their beauty and grace. As for the typographical execution of the volume, it will bear comparison with any English book, and quite surpasses most issues in America.—*Neal's Gazette.*

The admirers of the poet could not have his gems in a better form for holiday presents.—*W. Continent.*

# MISCELLANEOUS.

---

## ADVENTURES OF CAPTAIN SIMON SUGGS;

And other Sketches. By JOHNSON J. HOOPER. With Illustrations. 12mo, paper.......................................................50 cts.
Cloth.....................................................................62 cts.

---

## AUNT PATTY'S SCRAP-BAG.

By Mrs. CAROLINE LEE HENTZ, Author of "Linda." 12mo. Paper covers.......................................................50 cts.
Cloth.....................................................................62 cts.

---

## BIG BEAR OF ARKANSAS;

And other Western Sketches. Edited by W. T. PORTER. In One Volume, 12mo, paper........................................50 cts.
Cloth.....................................................................62 cts.

---

## COMIC BLACKSTONE.

By GILBERT ABBOT A' BECKET. Illustrated. Complete in One Volume. Cloth..........................................................75 cts.

---

## GHOST STORIES.

Illustrated by Designs by DARLEY. In One Volume, 12mo, paper covers.......................................................50 cts.

---

## MODERN CHIVALRY; OR, THE ADVENTURES OF CAPTAIN FARRAGO AND TEAGUE O'REGAN.

By H. H. BRACKENRIDGE. Second Edition since the Author's death. With a Biographical Notice, a Critical Disquisition on the Work, and Explanatory Notes. With Illustrations, from Original Designs by DARLEY. Two volumes, paper covers $1.00
Cloth or sheep..........................................................$1.25

## COMPLETE WORKS OF LORD BOLINGBROKE:

With a Life, prepared expressly for this Edition, containing Additional Information relative to his Personal and Public Character, selected from the best authorities. In Four Volumes, 8vo. Bound in cloth.........................................$6.00
In sheep.........................................................7.50

---

## CHRONICLES OF PINEVILLE.

By the Author of "Major Jones's Courtship." Illustrated by Darley. 12mo, paper..........................................50 cts.
Cloth............................................................62 cts.

---

## GILBERT GURNEY.

By Theodore Hook. With Illustrations. In One Volume, 8vo., paper....................................................50 cts.

---

## MEMOIRS OF THE GENERALS, COMMODORES, AND OTHER COMMANDERS,

Who distinguished themselves in the American Army and Navy, during the War of the Revolution, the War with France, that with Tripoli, and the War of 1812, and who were presented with Medals, by Congress, for their gallant services. By Thomas Wyatt, A. M., Author of "History of the Kings of France." Illustrated with Eighty-two Engravings from the Medals. 8vo.
Cloth gilt.......................................................$2.00
Half morocco.....................................................$2.50

---

## GEMS OF THE BRITISH POETS.

By S. C. Hall. In One Volume, 12mo., cloth............$1.00
Cloth, gilt......................................................$1.25

---

## VISITS TO REMARKABLE PLACES:

Old Halls, Battle Fields, and Scenes Illustrative of striking passages in English History and Poetry. By William Howitt. In Two Volumes, 8vo, cloth.....................................$3.50

## NARRATIVE OF THE ARCTIC LAND EXPEDITION.

By Captain Back, R. N. In One Volume, 8vo, boards...$1.50

## THE MISCELLANEOUS WORKS OF WILLIAM HAZLITT.

Including Table-talk; Opinions of Books, Men, and Things; Lectures on Dramatic Literature of the Age of Elizabeth; Lectures on the English Comic Writers; The Spirit of the Age, or Contemporary Portraits. Five Volumes, 12mo., cloth......$5.00
Half calf........................................................$6.25

## FLORAL OFFERING.

A Token of Friendship. Edited by Frances S. Osgood. Illustrated by 10 beautiful Bouquets of Flowers. In One Volume, 4to, muslin, gilt edges........................................$3.50
Turkey morocco super extra.....................................$5.50

## THE HISTORICAL ESSAYS,

Published under the title of "Dix Ans D'Etude Historique," and Narratives of the Merovingian Era; or, Scenes in the Sixth Century. With an Autobiographical Preface. By Augustus Thierry, Author of the "History of the Conquest of England by the Normans." 8vo., paper..................................75 cts.
Cloth ..........................................................$1.00

## BOOK OF THE SEASONS;

Or, The Calendar of Nature. By William Howitt. One Volume, 12mo, cloth..............................................$1
Calf extra.........................................................$2

## PICKINGS FROM THE "PORTFOLIO OF THE REPORTER OF THE NEW ORLEANS PICAYUNE."

Comprising Sketches of the Eastern Yankee, the Western Hoosier, and such others as make up Society in the great Metropolis of the South. With Designs by Darley. 18mo., paper..........................................................50 cts.
Cloth...........................................................62 cts.

## NOTES OF A TRAVELLER

On the Social and Political State of France, Prussia, Switzerland, Italy, and other parts of Europe, during the present Century. By SAMUEL LAING. In One Volume, 8vo., cloth.........$1

---

## HISTORY OF THE CAPTIVITY OF NAPOLEON AT ST. HELENA.

By GENERAL COUNT MONTHOLON, the Emperor's Companion in Exile and Testamentary Executor. One Volume, 8vo., cloth, $2.50
Half morocco.....................................................$3.00

---

## MY SHOOTING BOX.

By FRANK FORRESTER, (Henry Wm. Herbert, Esq.,) Author of "Warwick Woodlands," &c. With Illustrations, by DARLEY.
One Volume, 12mo., cloth......................................62 cts.
Paper covers.....................................................50 cts.

---

## MYSTERIES OF THE BACKWOODS:

Or, Sketches of the South-west—including Character, Scenery, and Rural Sports. By T. B. THORPE, Author of "Tom Owen, the Bee-Hunter," &c. Illustrated by DARLEY. 12mo, cloth, 62 cts.
Paper........................................................... 50 cts.

---

## NARRATIVE OF THE LATE EXPEDITION TO THE DEAD SEA.

From a Diary by one of the Party. Edited by EDWARD P. MONTAGUE. 12mo, cloth..........................................$1

---

## MY DREAMS:

A Collection of Poems. By MRS. LOUISA S. MCCORD. 12mo, boards ..........................................................75 cts

---

## AMERICAN COMEDIES.

By JAMES K. PAULDING and WM. IRVING PAULDING. One Volume, 16mo, boards...............................................50 cts.

## RAMBLES IN YUCATAN;

Or, Notes of Travel through the Peninsula: including a Visit to the Remarkable Ruins of Chi-chen, Kabah, Zayi, and Uxmal. With numerous Illustrations. By B. M. Norman. Seventh Edition. In One Volume, octavo, cloth.................................$2

---

## THE AMERICAN IN PARIS.

By John Sanderson. A New Edition. In Two Volumes, 12mo, cloth..........................................................$1

This is the most animated, graceful, and intelligent sketch of French manners, or any other, that we have had for these twenty years.—*London Monthly Magazine.*

---

## ROBINSON CRUSOE.

A Complete Edition, with Six Illustrations. One Volume, 8vo, paper covers..............................................$1.00
Cloth, gilt edges................................................$1.25

---

## SCENES IN THE ROCKY MOUNTAINS,

And in Oregon, California, New Mexico, Texas, and the Grand Prairies; or, Notes by the Way. By Rufus B. Sage. Second Edition. One Volume, 12mo, paper covers ..................50 cts.
With a Map, bound in cloth.......................................75 cts.

---

## THE PUBLIC MEN OF THE REVOLUTION:

Including Events from the Peace of 1783 to the Peace of 1815. In a Series of Letters. By the late Hon. Wm. Sullivan, LL. D. With a Biographical Sketch of the Author, by his son, John T. S. Sullivan. With a Portrait. In One Volume, 8vo, cloth...$2

---

## ACHIEVEMENTS OF THE KNIGHTS OF MALTA.

By Alexander Sutherland. In One Volume, 16mo, cloth, $1.00
Paper..........................................................75 cts.

---

## ATALANTIS.

A Poem. By William Gilmore Simms. 12mo, boards, 37 cts.

## LIVES OF MEN OF LETTERS AND SCIENCE.

By Henry Lord Brougham. Two Volumes, 12mo, cloth, $1.50
Paper ........................................................................$1.00

## THE LIFE, LETTERS, AND JOURNALS OF LORD BYRON.

By Thomas Moore. Two Volumes, 12mo, cloth..............$2

## THE BOWL OF PUNCH.

Illustrated by Numerous Plates. 12mo, paper...........50 cts.

## CHILDREN IN THE WOOD.

Illustrated by Harvey. 12mo, cloth, gilt..................50 cts.
Paper ........................................................................25 cts.

## CHARCOAL SKETCHES.

By Joseph C. Neal. With Illustrations. 12mo, paper, 25 cts.

## THE POEMS OF C. P. CRANCH.

In One Volume, 12mo, boards................................37 cts.

## THE WORKS OF BENJ. DISRAELI.

Two Volumes, 8vo, cloth ..........................................$2
Paper covers.................................................................$1

## NATURE DISPLAYED IN HER MODE OF TEACHING FRENCH.

By N. G. Dufief. Two Volumes, 8vo, boards..................$5

## NATURE DISPLAYED IN HER MODE OF TEACHING SPANISH.

By N. G. Dufief. In Two Volumes, 8vo, boards..............$7

## FRENCH AND ENGLISH DICTIONARY.

By N. G. Dufief. In One Volume, 8vo, sheep..................$5

## FROISSART BALLADS AND OTHER POEMS.

By Philip Pendleton Cooke. In One Volume, 12mo, boards..........................................................50 cts.

## THE LIFE OF RICHARD THE THIRD.

By Miss Halsted. In One Volume, 8vo, cloth...........$1.50

## THE LIFE OF NAPOLEON BONAPARTE.

By William Hazlitt. In Three Volumes, 12mo, cloth......$3
Half calf..............................................................$4

## TRAVELS IN GERMANY, BY W. HOWITT. EYRE'S NARRATIVE. BURNE'S CABOOL.

In One Volume, 8vo, cloth......................................$1.25

## STUDENT-LIFE IN GERMANY.

By William Howitt. In One Volume, 8vo, cloth...........$2

## IMAGE OF HIS FATHER.

By Mayhew. Complete in One Volume, 8vo, paper....25 cts.

## SPECIMENS OF THE BRITISH CRITICS.

By Christopher North (Professor Wilson). 12mo, cloth.75 cts.

## A TOUR TO THE RIVER SAUGENAY, IN LOWER CANADA.

By Charles Lanman. In One Volume, 16mo, cloth....62 cts
Paper.......................................................... 50 cts

## TRAVELS IN AUSTRIA, RUSSIA, SCOTLAND, ENGLAND AND WALES.

By J. G. KOHL. One Volume, 8vo. cloth..................$1.25

## LIFE OF OLIVER GOLDSMITH.

By JAMES PRIOR. In One Volume, 8vo, boards.............. .2

## OUR ARMY AT MONTEREY.

By T. B. THORPE. 16mo, cloth...............................62 cts.
Paper covers...................................................50 cts.

## OUR ARMY ON THE RIO GRANDE.

By T. B. THORPE. 16mo, cloth...............................62 cts.
Paper covers...................................................50 cts.

## LIFE OF LORENZO DE MEDICI.

By WILLIAM ROSCOE. In Two Volumes, 8vo, cloth..........$3

## MISCELLANEOUS ESSAYS OF SIR WALTER SCOTT.

In Three Volumes, 12mo, cloth..............................$3.50
Half morocco....................................................$4.25

## SERMON ON THE MOUNT.

Illuminated. Boards...........................................$1.50
" Silk...............................................................$2.00
" Morocco super.............................................$3.00

## MISCELLANEOUS ESSAYS OF THE REV. SYDNEY SMITH.

In Three Volumes, 12mo, cloth..............................$3.50
Half morocco....................................................$4.25

MRS. CAUDLE'S CURTAIN LECTURES...............12½ cts.

## SERMONS BY THE REV. SYDNEY SMITH.

One Volume, 12mo, cloth......................................75 cts.

## MISCELLANEOUS ESSAYS OF SIR JAMES STEPHEN.

One Volume, 12mo, cloth......................................$1.25

## THREE HOURS; OR, THE VIGIL OF LOVE.

A Volume of Poems. By MRS. HALE. 18mo, boards...75 cts

## TORLOGH O'BRIEN:

A Tale of the Wars of King James. 8vo, paper covers 12½ cts.
Illustrated......................................................37½ cts.

## AN AUTHOR'S MIND.

Edited by M. F. TUPPER. One Volume, 16mo, cloth....62 cts.
Paper covers......................................................50 cts.

## HISTORY OF THE ANGLO-SAXONS.

By SHARON TURNER. Two Volumes, 8vo, cloth...........$4.50

## PROSE WORKS OF N. PARKER WILLIS.

In One Volume, 8vo, 800 pp., cloth, gilt.....................$3.00
Cloth extra, gilt edges.............................................$3.50
Library sheep.......................................................$3.50
Turkey morocco backs.............................................$3.75
" extra.............................................$5.50

## MISCELLANEOUS ESSAYS OF PROF. WILSON.

Three Volumes, 12mo, cloth.......................................$3.50

## WORD TO WOMAN.

By CAROLINE FRY. 12mo, cloth...............................60 cts.

## WYATT'S HISTORY OF THE KINGS OF FRANCE.

Illustrated by 72 Portraits. One Volume, 16mo, cloth...$1.00
Cloth, extra gilt......................................................$1.25

www.ingramcontent.com/pod-product-compliance
Lightning Source LLC
LaVergne TN
LVHW020212110826
845151LV00003B/688

* 9 7 8 1 4 2 5 5 3 4 5 3 0 *